STEPPING AWAY FROM THE SILOS

CHANDOS
ADVANCES IN INFORMATION SERIES

Series Editors: David Baker
(Email: d.baker152@btinternet.com)
Wendy Evans
(Email: wevans@marjon.ac.uk)

Chandos is pleased to publish this major Series of books entitled Chandos Advances in Information. The Series editors are Professor David Baker, Professor Emeritus, and Wendy Evans, Head of Library at the University of St Mark & St John.

The series focuses on major areas of activity and interest in the field of Internet-based library and information provision. The Series is aimed at an international market of academics and professionals involved in digital provision, library developments and digital collections and services. The books have been specially commissioned from leading authors in the field.

New authors - we would be delighted to hear from you if you have an idea for a book. We are interested in short practically orientated publications (45,000+ words) and longer theoretical monographs (75,000–100,000 words). Our books can be single, joint or multi author volumes. If you have an idea for a book please contact the publishers or the Series Editors: Professor David Baker (d.baker152@btinternet.com) and Wendy Evans (wevans@marjon.ac.uk)

STEPPING AWAY FROM THE SILOS

Strategic Collaboration in Digitisation

MARGARET COUTTS

AMSTERDAM • BOSTON • HEIDELBERG • LONDON
NEW YORK • OXFORD • PARIS • SAN DIEGO
SAN FRANCISCO • SINGAPORE • SYDNEY • TOKYO
Chandos Publishing is an imprint of Elsevier

Chandos Publishing is an imprint of Elsevier
50 Hampshire Street, 5th Floor, Cambridge, MA 02139, United States
The Boulevard, Langford Lane, Kidlington, OX5 1GB, United Kingdom

Notices
Knowledge and best practice in this field are constantly changing. As new research and experience broaden our understanding, changes in research methods, professional practices, or medical treatment may become necessary.

Practitioners and researchers must always rely on their own experience and knowledge in evaluating and using any information, methods, compounds, or experiments described herein. In using such information or methods they should be mindful of their own safety and the safety of others, including parties for whom they have a professional responsibility.

To the fullest extent of the law, neither the Publisher nor the authors, contributors, or editors, assume any liability for any injury and/or damage to persons or property as a matter of products liability, negligence or otherwise, or from any use or operation of any methods, products, instructions, or ideas contained in the material herein.

ISBN: 978-0-08-100278-0 (print)
ISBN: 978-0-08-100279-7 (online)

Library of Congress Cataloging-in-Publication Data
A catalog record for this book is available from the Library of Congress

British Library Cataloguing-in-Publication Data
A catalogue record for this book is available from the British Library

For information on all Chandos Publishing publications
visit our website at https://www.elsevier.com/

Working together
to grow libraries in
developing countries

www.elsevier.com • www.bookaid.org

Publisher: Glyn Jones
Acquisition Editor: Glyn Jones
Editorial Project Manager: Lindsay Lawrence
Production Project Manager: Debasish Ghosh
Designer: Maria Ines Cruz

Typeset by TNQ Books and Journals

CONTENTS

BIOGRAPHY

Margaret Coutts was Chair of the Jisc Content Advisory Group and of the Jorum Steering Group from 2011 to 2015, following her retirement as University Librarian and Keeper of the Brotherton Collection at the University of Leeds. Over a period of some 35 years, she has also held posts at the universities of Glasgow, Aberdeen and Kent. Her experience spans strategic planning for and delivery of information services to all major academic disciplines, and includes the creation of a converged service comprising library, computing services, administrative computing and academic teaching support. In the UK and beyond, she has taken pivotal roles in major partnership ventures between universities, across wider education and heritage sectors, and between universities and top commercial firms. Her active involvement in professional organisations, including RLUK, Jisc, SCONUL and CILIP, has covered an extensive range of issues, from innovative collections and services to key skills and workforce development, with major concentration over many years on the development of the digital environment, both in the UK and Europe.

FOREWORD AND ACKNOWLEDGMENTS

This book has its origins in my experience as a Director of two university libraries, as an active member of various Jisc, RLUK and SCONUL groups, and in particular as Chair of the Jisc Content Advisory Group from 2011 to 2014. In all these capacities, I observed that the problematic issue of effectively coordinating digitised content at national level recurred frequently, and always remained unresolved. The purpose of this book is to take stock of the UK's digitisation efforts since the 1990s, and to consider whether a solution could be found in using the country's existing outputs from publicly and philanthropically funded initiatives as a basis for a national digitisation framework.

In gathering and assessing the evidence for this, I have been immensely grateful to the many individuals who provided me with information about digitisation in their fields, and whose advice helped to shape the ideas put forward here. My sincere thanks go to Valentina Asciutti, Chris Batt, Margaret Buchan, Luis Carrasquiero, Neil Curtis, Kim Downie, Crestina Forcina, Michelle Gait, Laura Gibson, Catherine Grout, Nick Hiley, Natalie Jones, Ian Lyne, John MacColl, Mike Mertens, Elspeth Millar, Francis Muzzu, Nick Poole, David Prosser, Laragh Quinney, John Scally, John Simmons, Mary Smith, Martyn Wade and Matthew Wheeler. I would also like to note the very significant time and effort contributed by Caroline Brazier, Andrew Green, Paola Marchionni, Simon Tanner and Phil Sykes. Finally, I owe much to the library staff teams at the universities of Sheffield, Glasgow and Aberdeen, at the National Library of Scotland and at Robert Gordon University, as well as the enquiry staff at CILIP. Their unfailing support has been exemplary at all stages of the research on which this book is based.

COVER ILLUSTRATIONS

John McArthur, *Plan of the city of Glasgow: Gorbells and Caltoun*, Glasgow, 1778.
Reproduced by kind permission of the National Library of Scotland.

Andreas Vesalius, *De humani corporis fabrica libri septem*, Basileae, 1543, p174.
Reproduced by kind permission of the Wellcome Library, London.

The Aberdeen Bestiary, c1200, Aberdeen University Library MS 24, f9r.
Reproduced by kind permission of the University of Aberdeen.

Michael Cummings, 'There is no alternative to me', *Sunday Express*, 2 Oct. 1988. Copyright Express Newspapers.
Reproduced by kind permission of N & S Syndication.
Image kindly provided by the British Cartoon Archive, University of Kent.

EDITORIAL NOTES

Throughout this book, references supporting quotations and specific facts include page numbers where these are available in the original publication but not where the cited resources are online documents without pagination.

The titles of information resources and services, whether online or analogue, are italicised throughout the text. The titles of digitisation programmes and projects are not italicised.

All web links in references were correct up to October 2015.

In the bibliographic references, UK, Northern Irish, Scottish and Welsh government publications are listed directly under the title of the issuing department. Official publications from other nations are listed under the name of the country.

Throughout the text, the form of 'Jisc' is used for both the current and pre-current versions of this organisation's title.

GLOSSARY

ACE   Arts Council for England
ACRL   Association of College and Research Libraries
AHDS   Arts and Humanities Data Service
AHRB   Arts and Humanities Research Board
AHRC   Arts and Humanities Research Council
BBC   British Broadcasting Corporation
Becta   British Educational and Communications Technology Agency
BIS   Department for Business, Innovation and Skills (United Kingdom)
BnF   Bibliothèque nationale de France
CEDARS   CURL Exemplars in Digital Archives
CNC   Centre national du cinéma et de l'image animée (France)
CURL   Consortium of University and Research Libraries
CyMAL   Museums Archives and Libraries Wales
DCAL   Department of Culture, Arts and Leisure (Northern Ireland)
DCMS   Department for Culture, Media & Sport (United Kingdom)
DELNI   Department for Employment and Learning (Northern Ireland)
DfES   Department for Education and Skills (the United Kingdom)
DIAD   Digitisation in Art and Design
DNER   Distributed National Electronic Resource
DPC   Digital Preservation Coalition
DPLA   Digital Public Library of America
DPN   Département des programmes numériques (France)
DTI   Department of Trade & Industry (United Kingdom)
EEVL   Edinburgh Engineering Virtual Library
eLib   Electronic Libraries Programme
EThOS   Electronic Theses Online Service
EU   European Union
FE   Further Education
FIGIT   Follett Implementation Group on Information Technology
GLAM   Galleries, libraries, archives and museums
GROS   General Register Office for Scotland
HATII   Humanities Advanced Technology and Information Institute, University of Glasgow
HE   Higher Education
HEA   Higher Education Academy
HEDS   Higher Education Data Sharing Consortium
HEFCE   Higher Education Funding Council for England
HEFCW   Higher Education Funding Council for Wales
HEI   Higher Education Institution
HELIX   Higher Education Library for Image Exchange
HL   House of Lords (United Kingdom)
HLF   Heritage Lottery Fund
HM Government   Her Majesty's Government (United Kingdom)

ICT   Information and Communications Technology
IfA   Initiatives for Access (British Library)
IFLA   International Federation of Library Associations and Institutions
IHR   Institute of Historical Research
Ina   Institut national de l'audiovisuel (France)
Jisc (formerly JISC)   Joint Information Systems Committee
MIDRIB   Medical Images: Digitised Reference Information Bank
MLA   Museums, Libraries and Archives Council
MoU   Memorandum of Understanding
NAO   National Audit Office
NAS   National Archives of Scotland
NEDCC   Northeast Document Conservation Center
NFF   Non-formula Funding of Specialised Research Collections in the Humanities
NI   Northern Ireland
NINCH   National Initiative for a Networked Cultural Heritage
NISO   National Information Standards Organization
NLM   National Library of Medicine (United States of America)
NLS   National Library of Scotland
NLW   National Library of Wales
NOF   New Opportunities Fund
NPO   National Preservation Office
NRS   National Records of Scotland
OER   Open Educational Resource
PRONI   Public Record Office of Northern Ireland
RAC   Register Archives Conversion Project
RCAHMS   Royal Commission on the Ancient and Historical Monuments of Scotland
RCUK   Research Councils UK
RDN   Resource Discovery Network
RLG   Research Libraries Group (United States of America)
RLS   Resources for Learning in Scotland
RLUK   Research Libraries UK
RMN   Réunion des musées nationaux (France)
RRMH   Research Resources in Medical History (Wellcome Trust)
RSLG   Research Support Libraries Group
RSLP   Research Support Libraries Programme
RUDI   Resources for Urban Design Information
SAfS   Scottish Archives for Schools
SCA   Strategic Content Alliance
SCAN   Scottish Archive Network
SFC   Scottish Funding Council
SLIC   Scottish Library and Information Council
SRG   Standard Research Grants (AHRB/C)
TARA   Trust for African Rock Art
TNA   The National Archives
UNESCO   United Nations Educational, Scientific and Cultural Organization

CHAPTER 1

Introduction: Digitisation since the 1990s

Some twenty years ago, digitisation rose to prominence in the educational, cultural and heritage sectors, bringing the prospect of revolutionised access to all forms of information and artefacts. Since then, it has gone through a cycle in which it has moved from high favour and priority to a more modest role, overshadowed by other developments in the digital world. It remains, however, a key element in the range of digital content on which all sectors now rely, and a key factor underpinning the current concepts, variously defined, of the 'digital environment' and the 'digital library' (Van Oudenaeren, 2010). As the digital world continues to present innovations that overshadow or replace earlier developments, it is timely to review digitisation progress to date, and to consider core aspects of the work that have had less attention than others as this groundbreaking development has come to maturity.

1.1 INITIATIVE AND INNOVATION

It is important to set such considerations in the context of the key developmental stages of digitisation. Terras (2011) includes in her full account the pre-1990 developments which created the basis for its widespread adoption from the 1990s onwards. Amongst the early adopters in the 1990s were the information and heritage sectors. They saw it as an unprecedented opportunity to extend access to their resources, to improve preservation, and to do so free of charge or at low cost. There was active support from leading educational and heritage organisations and significant amounts of public funding were invested. Total sums are very difficult to establish, but, for example, some £130 million was known to have been spent in the United Kingdom during the ten years up to 2005 (Bültmann et al., 2005, p. 3). All recognised that this was a true innovation. Lee described it as the 'decade of digitisation' (Lee, 2002, p. 160) and Lynch wrote of 'an enormous, exhilarating flowering of innovation, creativity and experimentation' (Lynch, 2000).

Stepping Away from the Silos
ISBN 978-0-08-100278-0
http://dx.doi.org/10.1016/B978-0-08-100278-0.00001-5

An entirely new field of expertise had to be developed, by a process of invention, experimentation, implementation, evaluation, refinement and further development. It also became clear that it was a multifaceted field, requiring that same process to be applied to issues of content, technology, infrastructure, intellectual property and sustainability. Universities, museums, galleries and national libraries were amongst the enthusiastic participants. In terms of content, small-scale projects typified the early work, often showcasing items of major intellectual and cultural value. The activity in this period, however, was as much concentrated on developing experience in and standards for the use of the technology and the provision of infrastructure, legal management and preservation. With standards and good practice consolidating around 2000, large-scale projects became more commonplace in the following years.

1.2 EXPANSION, CONSOLIDATION AND REVIEW

The wide variety of content produced in this period was welcomed by their communities, and user demand and expectation rose fast and high. By the 2000s, however, reality had also impacted. Digitisation was not a one-off, cheap solution, whether to enable access or to preserve, but a high-cost activity in terms of capital investment and, crucially, of recurrent expenditure. Most outputs required assured availability, periodic updating and long-term sustainability, all of which brought technical and financial implications. Moreover, the volume of material digitised to date was impressive, but constituted only a tiny fraction of the analogue resources yet to be considered for transfer to the digital environment. As Carr noted, 'the sums already spent from public and private sources have merely scratched the surface' (Carr, 2007, p. 49). There was also a 'fashion factor', summarised aptly by Van Oudenaeren. 'Paradoxically, efforts to digitise and place online additional content may be hampered somewhat by the very inexorability of the trends underway. Under the assumption that 'everything is going to be online eventually anyway', foundations, government agencies, and corporations may be less interested in supporting digitisation efforts than they were, for example, in the 1990s, when digitisation was a newer and more glamorous technology' (Van Oudenaeren, 2010, p. 98).

Some funders withdrew from supporting digitisation. Others became more cautious and judicious about their commitments and generally exercised a more strategic oversight of the work which they funded. This shift was intensified by the global economic downturn of 2008. Digitisation initiatives continued, but many reverted to small-scale and short-term projects.

Faced with such complications and setbacks, the enthusiasm and energy of the early days of digitisation have inevitably been replaced by a more subdued approach to continuing activity. This should not, however, obscure the widespread activity in the education and cultural sectors. A recent survey of digitisation of heritage materials in Europe revealed that, of some 2000 institutions responding, 83% had a digital collection or were currently involved in digitisation activities, and 34% had a written digitisation strategy (Stroeker and Vogels, 2012).

1.3 'A CORE FUNCTION'

Digital content in its widest sense is now seen, at the very least, as having a central role in key provision for education and heritage. A recent consultation with UK library directors drew the advice that 'digitised materials are of pervasive importance for many areas of Higher Education institutions' mission' (Marchionni, 2014). Terras describes the commitment of libraries, archives, museums, galleries and private collections to producing digital surrogates as 'commonplace' (Terras, 2011, p. 16). More significantly, there is strong evidence that digitised content is now of central importance in some fields and disciplines. In 2008, an environmental scan from the Association of College and Research Libraries (ACRL) Research Committee listed ten assumptions for forward planning, the first of which predicted that 'there will be increased emphasis on digitising collections, [and] preserving digital collections' (ACRL Research Committee, 2008). Broady-Preston and Swain (2012) quoted a member of the staff from the National Library of Wales (NLW), 'now [digitisation] seems a core function in giving access'. Hernon and Matthews observe that in some cases faculty are increasingly dependent on digital materials, including digital archives (Hernon and Matthews, 2012, pp. 3–4). Van Oudenaeren uses stronger terms still: various user groups display an 'increasing to near-total reliance' on electronic information, including cultural information (Van Oudenaeren, 2010, p. 1), and Calhoun notes several major surveys and studies recording calls for online provision (Calhoun, 2014, pp. 112–3).

This trend towards digital dominance is, as one would expect, influenced by users' needs and expectations. Stubbings (2012) notes an Ithaka survey which shows strong recognition of users' preference for online access and Calhoun stresses that 'most segments of population place high value on immediately available, convenient online sources' (Calhoun, 2014, p. 112). Coyne considers convenience for the academic sector specifically, and believes that students, researchers, faculty and authors need to access information around the clock and from anywhere in the world (Coyne, 2010, pp. 104–5). Significantly, many commentators are signalling that we may well be approaching, finally, the long-predicted time when digital supersedes analogue in standard provision across all fields of interest:

> We can assume that most printed out-of-copyright paper publications will be digitised and fully available on the web by 2030
> **(National Library of Scotland, 2010,** p. 7**)**

> There is a growing expectation in society that all information resources should be available online 24 hours a day.
> **(The National Archives and Museums, Libraries and Archives Council, 2010,** p. 6**)**

> Digital formats are beginning to dominate library collections, especially in academic libraries.
> **(Calhoun, 2014,** p. 111**)**

> ...the digital versions will be the principal mode of delivery
> **(Pressler, 2014,** p. 12**).**

This trend is, of course, enabled by the pervasive nature of the digital environment and the ubiquity of born-digital materials that have long replaced physical formats in many fields. Over the last 10 years or so, this has obscured to a certain extent the case for progressing further digitisation of analogue materials. In the light of increasing digital dominance, however, and the incontrovertible fact that so much of the world's knowledge and cultural outputs still exist only in physical form, digitisation is as much a key element in future digital developments as it has been in the past. 'Digitization of existing library, museum, and archive collections is still a major priority, where funding can be found for these initiatives' (Hughes, 2012, p. 1).

1.4 CURRENT ISSUES

There are considerable risks in not advancing digitisation at this time. The numerous 'digital-only' users may ignore or never know of key content which is relevant to their interests and needs. Neglected knowledge, over time, becomes lost knowledge, to the detriment of all. The corollary is the likelihood that knowledge which already exists in analogue form will be created afresh in digital form, wasting intellectual and financial resources.

Digitisation in these times has the benefit of the sound experience built up over the last 20 years. The digital lifecycle is now well defined, moving from creation through curation, preservation, discovery and use to the creation of new knowledge and content. Standards and good practice for each of these elements have been systematically tackled, particularly in the early years of digitisation (Dunning and Marchionni, 2011), with procedures and technical guidelines being well documented to support further developments (Terras, 2011, p. 14). New initiatives will have a solid base of guidance in matters of technical specifications, infrastructure, sustainability and intellectual property. While there remain many complexities in these fields of expertise, they benefit from continual updating; further digitisation work will in turn contribute to their development. Similarly, skills required by the staff to apply the technology and standards are now well established and extended as new technologies and methods develop.

A crucial advantage for further digitisation is a level of realism in the vision, conception and planning of initiatives, informed by the established knowledge and practices described in the previous paragraph. It can be argued, however, that the necessity of researching and resolving all these issues to deliver the digitised materials, and the sheer scale of this work over many years, has resulted in less consistent attention to other aspects of digitisation work.

1.5 A SILO CULTURE

Perhaps the most marked of these issues is a widespread lack of coordination of digitisation initiatives, both within and across different sectors. The literature and, indeed, professional conversations abound with observations lamenting this weakness:

> Funding of the creation of digital content in the UK and around the world has been piecemeal and completely uncoordinated
> **(Bültmann et al., 2005,** p. 3**)**

> 'There is not…much evidence…of co-ordination either within the HE/FE sector or across the public and third sectors'
> **(chrisbatt Consulting, 2009,** p. 45**)**

> The result has been to create digital silos — lumps of digital content that cannot be shared, reused or cross-searched without considerable difficulty, even when their content is of a similar nature
> **(Dunning, 2010,** p. 123**)**

'It is a major concern that the process of digitization is somewhat random, with a large number of players making independent decisions, leading to a fragmented picture of national and international provision' (House, 2012, p. 57).

It is unfair to say that there has been no effort at all in this area. Various funding programmes have required coordination of content, technology and skills, as have many strong collaborations and partnerships over the years. Nonetheless most attempts have proved limited in scope and outcome so far.

1.6 CONTENT SELECTION

There has been less explicit consideration over the years of the selection of content for digitisation. Different initiatives have set different criteria for choosing content, and these have varied from the most general to the highly targeted and specific. There are justifiable reasons for these approaches, especially in the early years when experimentation and development of standards necessarily dominated. Nonetheless, this approach has provided users with a patchwork of digitised materials where there are many identified gaps, and there is also some degree of duplication (Pearson, 2001). Unsurprisingly, this latter concern has been raised in relation to funding issues in particular; the National Audit Office, for example, noted such concerns over 10 years ago (National Audit Office, 2004).

Overall, if these issues remain unresolved, they will continue to limit the benefits and impact of the digitised materials produced, and of the financial and human investments which they represent. This book will examine the strategic context in the United Kingdom since the 1990s and how it has affected collaboration and coordination of exemplar initiatives in the educational and cultural sectors to date. It will identify the core international literature on criteria for the selection of intellectual content and the principal criteria that are common to the different publications. The outputs of the exemplar projects will be examined in relation to these criteria. This will lead on to consideration of how far practices and patterns in content selection may have developed over the years relating to these criteria, and whether they could contribute to improved coordination and to addressing the vexed issue of a national digitisation strategy for the United Kingdom. Other key elements of digitisation which are already widely covered in the literature, including technical specifications, network and infrastructural matters, preservation, sustainability and intellectual property, will be considered only where they are directly relevant to the above issues.

1.7 SCOPE AND DEFINITIONS

This book will consider these issues primarily from the perspective of higher education in the United Kingdom. It will, however, examine activity in the wider educational and heritage sectors in the United Kingdom. This will include work in public libraries, archives and museums. Digitisation is, of course, a global activity, and the study will also be set in this broader context, particularly in relation to issues of collaboration and national digitisation strategies. It will concentrate on initiatives financed primarily from public or philanthropic funds, sources which generally promote open access and are free of commercial requirements and restrictions. Commercially funded initiatives will, however, be covered where relevant, typically in the case of public/private partnerships and in instances where the content is accessible to the public without charge in identifiable locations.

For the purposes of this book, 'digitisation' is taken to mean 'digital content creation by making a digital copy or digital recording of analogue information, where that information can reside in a document, artefact, sound, performance, geographical feature or natural phenomena' (DigitalNZ, 2009). The generic term 'digitised content' embraces many forms of digitised materials, some of which are already very well documented in the literature in relation to their development and the issues outlined previously. These include e-journals, e-books, research outputs, research data, and open educational resources (OERs). This book will concentrate on the wider and more amorphous range of primary materials beyond these categories. There are no clear boundaries, however, between different digital forms, and therefore these categories will also be considered where there is a direct relationship with the principal set of materials and the issues under consideration.

REFERENCES

ACRL Research Committee. (2008) *Environmental Scan 2007*. Chicago ACRL. Available. [Online] Available from: http://www.ala.org/sites/ala.org.acrl/files/content/publications/whitepapers/Environmental_Scan_2007%FINAL.pdf.

Broady-Preston, J. and Swain, W. (2012) What business are we in? Value added services, core business and national library performance. *Performance Measurement and Metrics*. Vol. 13, No. 2: 107–120.

Bültmann, B., Hardy, R., Muir, A. and Wictor, C. (2005) *Digitised Content in the UK Research Library and Archives Sector: a report to the Consortium of University and Research Libraries and the Joint Information Systems Committee*. [Bristol?]: JISC; CURL. [Online] Available from: *http://www.jisc.ac.uk/uploaded_documents/JISC-Digi-in-UK-FULL-v1-final.pdf*.

Calhoun, K. (2014) *Exploring Digital Libraries*. London: Facet.

Carr, R. (2007) *The Academic Research Library in a Decade of Change*. Oxford: Chandos. Information Professional Series.

chrisbatt Consulting. (2009) *Digitisation, Curation and Two-way Engagement: Final Report*. [Online] Available from: *http://www.jisc.ac.uk/media/documents/programmes/digitisation/dcatwefinalreport_final.pdf*.

Coyne, P. (2010) Loosely joined: the discovery and consumption of scholarly content in the digital era. In McKnight, S. (Ed.) (2010). *Envisioning Future Academic Library Services: initiatives, ideas and challenges*. London: Facet: 101–118.

DigitalNZ. (2009) *Selecting for Digitisation*. [Online] Available from: http://www.digitalnz.org/make-it-digital/selecting-for-digitisation.

Dunning, A. (2010) Digitizing the past: next steps for public sector digitization. In Woodward, H. and Estelle, L. (Eds.) *Digital Information: order or anarchy?* London: Facet: 117–131.

Dunning, A. and Marchionni, P. (2011) Introduction. In *JISC Clustering and Sustaining Digital Resources: the JISC eContent Programme 2009-2011*: 1–4. [Online] Available from: http://www.webarchive.org.uk/wayback/archive/20140615013612/http://www.jisc.ac.uk/media/documents/publications/general/2011/JISCeContentClusteringAndSustainingDigitalResources.pdf.

Hernon, P. and Matthews, J. R. (Eds.) (2012) *Reflecting on the Future of Academic and Public Libraries*. London: Facet.

House, D. (2012) An overview of e-resources in UK further and higher education. In Fieldhouse, M. and Marshall, A. (Eds.) *Collection Development in the Digital Age*. London: Facet: 47–58.

Hughes, L. M. (2012) Introduction: the value, use and impact of digital collections. In Hughes, L. M. (Ed.) *Evaluating and Measuring the Value, Use and Impact of Digital Collections*. London: Facet: 1–10.

Lee, S. (2002) *Digital Imaging: a practical handbook*. London, Facet.

Lynch, C. (2000) From automation to transformation. *Educause Review*. Vol. 35, No. 1: 60–68.

Marchionni, P. (2014) *Library Directors' Views on Digitisation*. [Online] Available from: http://digitisation.jiscinvolve.org/wp/2014/01/15/library-directors-views-on-digitised-collections/.

National Audit Office. (2004) *The British Library: providing services beyond the Reading Rooms*. London: Stationery Office. HC 879 (2003-2004). [Online] Available from https://www.nao.org.uk/report/the-british-library-providing-services-beyond-the-reading-rooms/.

National Library of Scotland. (2010) *Thriving or Surviving?: National Library of Scotland in 2030*. [Online] Available from: http://www.nls.uk/media/808985/future-national-libraries.pdf.

Pearson, D. (2001) Developing the Wellcome 'Digital Library'. *Wellcome History*. Vol. 18, No. 9. [Online] Available from: http://www.wellcome.ac.uk/stellent/groups/corporatesite/@msh_publishing_group/documents/web_document/wtd006085.pdf.

Pressler, C. (2014) *National Digitisation Review: Shifting Sands: RLUK Board Briefing Paper*. [Online] Available from: http://www.rluk.ac.uk/wp-content/uploads/2014/12/RLUK-National-Digitisation-Review-CPressler.pdf.

Stroeker, N. and Vogels, R. (2012) *Survey Report on Digitisation in European Cultural Heritage Institutions 2012*. ENUMERATE Thematic Network. [Online] Available from: *http://www.enumerate.eu/fileadmin/ENUMERATE/documents/ENUMERATE-Digitisation-Survey-2012.pdf*.

Stubbings, R. (2012) Supporting users to make effective use of the collection. In Fieldhouse, M. and Marshall, A. (Eds.) *Collection Development in the Digital Age*. London: Facet: 197–210.

Terras, M.M. (2011) The rise of digitization: an overview. In Rikowski, R. (Ed.) (2011) *Digitisation Perspectives*. Rotterdam: Sense Publishers: 3–20.

The National Archives and Museums, Libraries & Archives Council. (2010) *Archives for the 21st Century in Action*. [Online] Available from: http://www.nationalarchives.gov.uk/documents/information-management/archives-for-the-21st-century-in-action.pdf.

Van Oudenaeren, J. (2010) Strategies for institutions: responding to the digital challenge: the World Digital Library perspective. In Verheul, I., Tammaro, A.M. and Witt, S. (Eds.) *Digital Library Futures: User Perspectives and Institutional Strategies*. Berlin: De Gruyter-Saur: 97–105.

CHAPTER 2

Strategic Context

To examine strategic collaboration in digitisation, and especially in the selection of intellectual content, it is important to understand the wider strategic context in which such choices are set. The projects and organisations that are the exemplars for this book have all been subject to or influenced by the strategies of government and of governmental and sectoral agencies, as well as those of their own parent organisations. Therefore, the level of strategic commitment, or noncommitment, to digital content creation and digitisation is an important factor in the initiation and delivery of such content. This chapter summarises the strategies and related documents of the relevant bodies, and highlights where and how digital content and digitisation is positioned within them. Past strategies as well as current ones are examined where these reflect either significant changes or clear consistency of practice over time.

2.1 UK AND DEVOLVED GOVERNMENTS

Since the late 1990s, there has been a series of major strategies and policies which have presented governmental agendas for development and exploitation of the maturing digital environment (Department of Trade and Industry (DTI), 1998; Cabinet Office; Prime Minister's Strategy Unit and DTI, 2005; Department for Culture, Media & Sport (DCMS) and Department for Business, Innovation & Skills (BIS), 2009; HM Government, 2013; DCMS, 2013a; Innovate UK, 2015; Scottish Executive and One Scotland, 2006; Scottish Government, 2011b, 2012, 2013; Welsh Assembly Government, 2010, 2011a, 2013a, 2013b; Welsh Government, 2014a; Invest Northern Ireland, 2010; Digital Northern Ireland Advisory Board, 2010; *Northern Ireland Digital Content Strategy 2012–15*, c2011). While there are some differences in nuance between the central and devolved positions, they all reflect the same core perspective. They recognise the digital environment as firmly established and increasingly pervasive at economic and societal levels, and see this as the basis for a 'digital economy'. They all advocate maintaining a global position amongst the most advanced digital nations. Sustaining and strengthening this position is seen as key to future economic success and to wider social and

Stepping Away from the Silos
ISBN 978-0-08-100278-0
http://dx.doi.org/10.1016/B978-0-08-100278-0.00002-7

educational improvement. The emphasis throughout is strongly focused on the value of the digital environment to business and enterprise, and on social benefits in the widest sense. Within that, they present objectives for digital inclusion; digital delivery of public services to all or nearly all of the population; skills development; embedding digital delivery in all levels of education; infrastructure development, in particular pervasive, high-speed broadband; data capability and management; and online security.

It is self-evident that the creation of digital content in all its forms, whether born-digital or digitised, is and will be a core element in delivering these strategies, but explicit recognition of this is not consistent over the period which the documents cover, and they tend to focus on the commercial production models and economic benefits of such developments, especially at the UK level. In the earlier documents from Westminster, discussion of digital content and its role concentrates on the outputs of the BBC and other broadcasting media. All the documents emphasise the digital delivery of public information and services, but there is considerable variation in references to the means by which the relevant content, whether born digital or digitised from analogue, will be created. By and large, this has to be inferred from various sections of these documents.

It is in 2009, with the publication of *Digital Britain*, that the UK government most clearly acknowledges and embraces the role of digital content. As before, there is much emphasis on broadcasting outputs, but the strategy also recognises that the wider public sector, as well as the government, are major commissioners of digital content, and that cultural institutions of all types, including art galleries, theatres, opera houses, film councils, libraries, museums, and archives are reaching a wider public through their online activities. There is considerable optimism for the future of such ventures. 'We have a mixed economy of content creation, often taking the best from the worlds of subsidy and commerce…many of the basic building blocks in which people develop and enjoy creative content live — our museums, libraries, arts centres, theatres and music venues — look set to thrive in the Digital Age' (DCMS and BIS, 2009, p. 107). However, while there are passing references to publicly funded digital content, for example, the British Library's large-scale digitisation of newspapers and the role of the British Film Institute (DCMS and BIS, 2009, pp. 122 and 132, respectively), the strategy's financial perspective for the future, consistent with the earlier documents, is on business models with a commercial basis. The later manifestations of UK government thinking on digital strategy include very little relating to digital content in its widest role other than updated commentary on broadcasting outputs (HM Government, 2013; DCMS, 2013a; Innovate UK, 2015).

If these reports and policy documents make no explicit commitment to supporting digitisation as a key element of the overall strategy, a series of reports from the House of Lords Select Committee on Science and Technology goes some way to rectifying this (House of Lords; Science and Technology Committee, 2006, 2007; DCMS, 2012). While the Committee principally focuses on the development of 'Heritage Science' to address issues of digital preservation and sustainability, it does acknowledge clearly that 'the digitisation of records by libraries and archives is of particular importance…in making records more widely accessible and more readily searchable' (House of Lords; Science and Technology Committee, 2006; para 2.12) and 'the ability to digitise collections and artefacts is one of the key technological developments of recent times' (House of Lords; Science and Technology Committee, 2006; para 7.12). The principal outcome of this thinking was support for developing a National Heritage Science Strategy and this was underpinned by the detailed evidence submitted for the report from many organisations with experience in digitisation.

The strategies for Northern Ireland carry the same emphasis on the infrastructural, commercial and social benefits of the digital environment, and the attendant priorities for action within the province. The digital strategies and related documents of the Scottish and Welsh governments express more active support for born-digital and digitised content. A full-scale review of Scotland's cultural landscape in 2006 called for a brand new strategic approach to the delivery of electronic material, with digitisation explicitly mentioned as an area requiring research (Scottish Executive and One Scotland, 2006, pp. 44 and 53). In more recent years, *Scotland's Digital Future* made clear the importance of cultural content, albeit with emphasis on the value to tourism, and of digital technologies in the country's strong creative industries (Scottish Government, 2011b). It described in some detail major examples of digitisation work carried out by government-financed organisations, including the National Archives of Scotland (NAS) and the National Library of Scotland (NLS), and called for an overall strategy for data archiving to underpin such activities. In Wales, government strategies and policies have placed a very strong emphasis on the importance of cultural inclusion in many aspects of national development. In this context, digital content is seen as having a key role, and the contribution made by digitisation work in libraries, art galleries, museums and others across Wales to social inclusion, skills development and economic growth is explicitly and repeatedly highlighted (Welsh Assembly Government, 2010, 2011a, 2013a, 2013b; Welsh Government, 2014a).

2.2 HIGHER EDUCATION

The strategic significance of digital content and digitisation is amplified in the strategies of government agencies responsible for the various educational sectors within their remit.

2.2.1 Higher Education: Funding Councils

The overarching strategies and reports of the Higher Education Funding Council for England (HEFCE) recognise the use of new technologies for learning, teaching and research in very high-level terms. It is in the Council's strategies for e-learning, developed from 2005 to 2008, that the relevance of digital content is most explicit (HEFCE et al., 2005, HEFCE, 2009, 2011; HEFCE; Online Learning Task Force, 2011; Glenaffric, 2008). Its 2005 strategy includes the aim 'to ensure that there is confident use of the full range of pedagogic opportunities provided by ICT'. Its seven strands for development include one for learning resources and networked learning. Under this, Jisc and the Higher Education Academy (HEA) are charged with developing 'a comprehensive and coherent approach to the development and use of resources for learning and teaching, including digital resources and discovery tools'. The review of this strategy, carried out in 2008, shows a wide range of findings and outputs from Jisc and the HEA, including specific recognition of Jisc's Digitisation Programme (see Chapter 3) as a central plank in progressing the overall strategy (Becta, 2008).

If HEFCE's thinking at this time acknowledges the need for digital content creation, this is even more clearly articulated in the complementary strategies for the entire education sector, issued by the Department for Education and Skills (DfES) and Becta in the same period and in parallel to the HEFCE documents (DfES, 2005; Becta, 2008). Like HEFCE, the issuing bodies identify a range of developments to embed technology in all aspects of learning and teaching. Within this, the importance of digital content and digitisation is recognised throughout the strategies:

Many impressive digital library resources are being developed across education and through other government departments such as the Museums and Libraries Archives Council (sic)...and the e-Science resources (DfES, 2005, p. 26).

Digital resources to support learning and teaching are now increasingly available across sectors and borders…we will explore how the best resources can be made available. We also need to pool effort in the way such resources are procured…we need to promote national arrangements for the collaborative development of content and services, to enhance front-line value for money and reduce duplicated efforts (Becta, 2008, p. 32).

In more recent years, HEFCE's updated strategies have shown a consistent commitment to using new technology for the enhancement of learning and teaching, but with a clear shift to the development of tailored learning materials and of open educational resources (OERs) (HEFCE, 2009, 2011a, 2011b; Leadership Foundation et al., c2014). Jisc and HEA are again called on to instigate the research and promulgation of principles and practice. There is also a very clear emphasis on the responsibility of individual institutions to establish strategies appropriate to local circumstances, within the framework offered by the Funding Council.

The most recent strategies of the devolved nations' Funding Councils reflect the same growth of interest in open learning and OERs, albeit more briefly. In the case of the Scottish Funding Council (SFC), there has been a conscious shift towards expressing strategy and its delivery in terms of outcomes rather than mechanisms (Scottish Government, 2011a; Scottish Funding Council, c2011, c2012a, 2012b), and the likely digital developments are better reflected in the government documents discussed in the previous section of this chapter. The Welsh and Northern Irish strategies carry a particularly strong emphasis on widening access, as well as cultural and community engagement (HEFCW, 2010, 2011, 2013a, 2013b, 2013c; Department for Employment and Learning, Northern Ireland, c2012). It is implicit rather than explicit that the digital environment offers the means of achieving the vision presented for each administration.

2.2.2 Higher Education: Jisc

The exploitation of new technologies is only one element in the wider remits of the Funding Councils, and it is to be expected that their strategies will present only limited coverage of specific matters such as the role of digital content and the means of its creation. This is particularly justified because the Councils' long-standing and successful delegation to Jisc of ICT development and implementation has allowed digital issues to be explored in this separate arena, where the immense complexities of the field have been thoroughly investigated and effective solutions delivered.

Like the Funding Councils, Jisc strategies also cover the full range of higher education needs, and it is therefore significant that, while other issues come and go, digital content is consistently present from the very first initiatives of the 1990s to the present day. Over these 20 years, Jisc's coverage embraces born-digital and digitised outputs, produced by commercial suppliers as well as by public and philanthropic funding. The strategic variations are found in the level of priority given to different outputs in different periods. Recognising the early successes of the eLib programme (Jisc, 1996), the strategies for the first decade of the 21st century made a commitment to supporting further development of digital content, with clearly stated priorities. 'The JISC (sic) will pursue an active collections policy concentrating in the areas of electronic books, finding aids, geospatial resources, still and moving images, learning materials and primary research data' (Jisc, 2001, p. 16). This content was to be 'generated by others within the UK education and research community and from the private sector' (Jisc, 2001; para 17). As this policy developed, a major programme of large-scale digitisation was initiated (see Chapter 3). The experience gained gave digitisation prominence in the organisation's strategy for 2007−9 (Jisc, c2007), and led to the development of a formal Jisc Digitisation Strategy (Jisc, 2008). Later strategies continued to acknowledge the role of digitisation, but priorities shifted, for example to the development of OERs, a direct and necessary response to Funding Council requirements, and to the growing demand for effective management of research data (Jisc, c2010). In its most recent strategy and supporting documents, the first since Jisc became a legally independent organisation, there is a high-level commitment to innovating and improving the organisation's digitised portfolio and to developing digitised resources (Jisc, 2013, c2014).

2.3 RESEARCH COUNCILS

It is fair to say that the work of many, if not all, of the UK Research Councils depends increasingly on digital content, and frequently on key information that can be provided by digitising analogue content. Over the period covered by this book, the Research Councils have placed ever-increasing emphasis on research intended to address the most challenging and pressing of global issues. In this context, it is unsurprising that most of their strategies do not include acknowledgement of many activities, such as digitisation, which might be seen rather as enablers in achieving their goals. With the rapid development of Digital Humanities as a discipline in its own right, the exception is to be found in the perspective of the Arts and Humanities Research Council (AHRC). Its predecessor, the Arts and Humanities Research Board (AHRB), was established in 1998 and amongst its earliest initiatives was the Resource Enhancement Scheme, set up to 'make accessible resources and scholarly information in digital form' (Dunning, 2010, p. 119).

This programme ran from 2000 to 2006 (Denbo et al., 2008), after which further digitisation work was absorbed by different and wider-ranging programmes. At a strategic level, however, digitisation retains an explicit place in the Council's thinking. It can be seen as relevant to several sections of the Council's more recent strategies (AHRC, c2007) and in particular to the cross-Council theme of Digital Transformations (AHRC, 2013). More specifically, the Council's precurrent *Delivery Plan* referred directly to digitisation as a continuing element in further exploitation of digital technologies, although this was less prominent than other de-velopments that relate more directly to the intensifying focus on societal and global issues (AHRC, 2010). In the current *Delivery Plan*, these issues and key strategic priorities predominate, and the role of digital technology as a whole is an embedded factor in progressing the Council's vision, within which digitisation will be one of many methods to be used as appropriate (AHRC, 2010).

2.4 NATIONAL LIBRARIES

From around 2000 onwards, the strategies of the UK's three national libraries all include the development of their digital libraries as an essential element of their provision both in the immediate future and the long term. They recognise that the content will be provided from the increasing amount of born-digital material becoming available and also from digitisation of their existing analogue collections. There have been, however, variations in emphasis between the libraries' strategies which are worthy of note.

2.4.1 British Library

The British Library was amongst the most prominent of the first adopters of digitisation. As early as 1993, the Library's strategic objectives included the statement that 'by the year 2000, the British Library will be a major centre for storage of and access to digital texts required for research' and that 'by the year 2005, our digital collection will be enormous and growing at a huge rate' (Mahoney, 1998, p. 11). Brian Lang, the then Chief Executive, envisioned the digital library as an integral part of the way that the Library would fulfil its responsibilities in the future, and the many sources of such provision included digitisation of British Library content (Lang, 1998). This strategic position led to a range of early groundbreaking projects under the Library's Initiatives for Access programme (Carpenter et al., 1998).

By the year 2000, digital content and digitisation were explicit in the Library's strategy. In her new role as the Library's Chief Executive, Lynne (later Dame Lynne) Brindley stated that 'the British Library needs to turn to and accelerate its engagement with the digital world' (Brindley, 2000). Its strategic priorities included the improvement of access to collections through digitisation (British Library, 2001). The aim was to create a critical mass of digitised materials, and to do so with added value from commentaries, critical additions, scholarship and instruction. In the following years, digitisation became mainstream with the strategic emphasis on mass digitisation as part of the envisaged National Digital Library (British Library, 2005). As a new decade began, *Growing Knowledge* covered 2011–15 (British Library, 2011) and reflected the national and global economic pressures of the period. Many long-term developments were redefined or limited for financial reasons. It was therefore significant that, despite serious constraints in all areas of the Library's business, digitisation still held a prominent place in the key implementation priorities (British Library, 2011, p. 5). In support of this, the Library's separate Digitisation Strategy, implemented in 2008, remained in force, pursuing its energetic vision:

> We aim to help researchers advance knowledge by becoming a leading player in digitisation. We will produce a critical mass of digitised content, reflecting the breadth and depth of our collection. We will provide a compelling user experience that facilitates innovative methods of research and meets 21st century requirements for interacting with content (British Library, 2008).

In the Library's most recent strategy, *Living Knowledge*, digitisation retains its importance as a core method of delivering many of the major developments planned. It is also at the heart of the new commitment to follow the successful programme of newspaper digitisation with one of equally large scale which will expand digital access to the Library's major collections of rare and unique sound recordings (British Library, c2015).

2.4.2 National Library of Scotland

In the early years of the 21st century, the National Library of Scotland was renewing its overall strategy. The fully revised plan, *Breaking through the*

Walls, recognised the need for 'new ways to help users find and interpret [the Library's materials]' (NLS, 2003, p. 15). Its specific policies for delivering the new strategy included placing 'a high priority on extending our collection of electronic resources through digitisation' (NLS, 2003, p. 23). In its supporting strategy for creating the Digital National Library of Scotland, the need for digitisation permeated the entire document, as well as constituting a specific objective with an undertaking to scale up digitisation (NLS, 2005). In subsequent strategies and related planning documents, digitisation appears to have settled as a continuing strategic priority, with little variation in the level or type of activity to be taken forward. Instead, it is in the Library's corporate and action plans that the commitment to develop and extend digitisation is reaffirmed (NLS, 2008, 2011, 2012, 2013). The most recent of these documents at the time of writing, the *Corporate Plan* for 2013–14, includes an explicit commitment to a long-term programme for digitisation of analogue collections, alongside the development of a national strategy for digital access to Scotland's cultural heritage, and the aim to provide digital access to all Scottish out-of-copyright publications (NLS, 2013, pp. 10, 16 and 29).

2.4.3 National Library of Wales

During the 1990s, the National Library of Wales explored digitisation through exhibitions and a small number of pilot projects (Jones, 2008). Around the year 2000, the Library made such a strong strategic commitment to the concept of the digital library and to digitising its collections that its Vision Statement included specific reference to the latter. Its strategies over the following decade included objectives focussing directly on digitisation, and with each new version there was a noticeable rise in its importance. By 2008 it was clear to the Library that 'there is agreement that digitisation is an excellent means to open the Library's collection to a worldwide audience' (NLW, 2007, p. 20), and by 2011 there was strong commitment to carrying out yet more extensive digitisation and to delivering the established Digitisation Programme (NLW, 2011a). This level of engagement is continued in the Library's current strategy (NLW, 2014a). The consistency of commitment in overall strategy is underpinned by thorough and detailed digitisation strategies, the first of which was issued as early as 2000 (Green, 2002, p. 166) with the second following in 2005 (Jenkins, 2005). Building on the very small-scale outputs of the 1990s, these strategies steadily moved the Library from 'boutique' digitisation to large-scale, complex initiatives, aiming for critical mass in specific areas. By 2008 digitisation was 'at the heart of the National Library of Wales' strategy' (NLW, 2009, p. 5). This is further reinforced in the most recent digitisation strategy, covering 2011—15, where it is acknowledged that digitisation, having been initiated to improve access, is now also key to the Library's preservation activities (NLW, 2011b).

Digitisation has been well embedded in the national libraries' strategies up to the time of writing. This is especially true in the cases of the British Library and the National Library of Wales, where the value of digitisation is explicitly reinforced in each strategic phase. It should be noted at this point that the strategy of the National Library of Scotland was under revision at the time of writing this book, and therefore it has not been possible to establish the exact role of digitisation in the new plans. However, it is worth bearing in mind that in 2010 all three national libraries drew up longer term visions for their libraries. The British Library and the National Library of Wales considered the period up to 2020 (British Library, 2010; NLW, 2010), while the National Library of Scotland extended its thinking to 2030 (NLS, 2010). All these documents were based on extensive consultation with the libraries' stakeholders. They took full account of the political, financial and technological factors which would affect them, and assessed the impact of this wider environment on the libraries' future collections and services. These observations influenced directly the national libraries' strategies from 2011 onwards, and they can be expected to remain part of the information base which will contribute to future planning. Each document has its own distinctive character, reflecting the differences between the libraries in scale and in the communities which they serve. Nonetheless, there are some issues which are common to all, and these include the key role of digitisation. All make repeated reference to this element of the digital library, with strong emphasis on the need to continue digitising:

The key role we believe we can play in liberating access to public-domain content is through digitisation (British Library, 2010, p. 7).

There…is enormous scope for the National Library of Scotland to digitise unique material (NLS, 2010, p. 7).

In practice this will mean extending digitisation (NLW, 2010, p. 26).

With the British Library's most recent strategy (British Library, c2015), covering a planning period to 2023, the full influence of its forward vision document is now evident, with digitisation, as already noted, an explicit element of its plans. It seems reasonable to assume that digitisation will also figure in the new strategic priorities for the National Library of Scotland, as it has in the strategy to 2017 for the National Library of Wales (NLW, 2014a).

2.5 PUBLIC LIBRARIES

The strategic context for digital content creation, including digitisation, in the UK public library sector is set separately by each of the four devolved administrations, with consequent variations in issues and priorities. As a major review of UK public libraries for the Carnegie Trust UK noted, the 'political, economic and structural context of these library services are different and have become increasingly divergent as a result of devolution' (Macdonald, 2012, p. 3). The development of principal themes and trends in each of the jurisdictions can be found in a series of key documents from the Department for Culture, Media and Sport (DCMS, 2003, 2010, 2013a, 2013b, 2014a, 2014b), the House of Commons Culture, Media and Sport Committee (2005), the Museums, Libraries and Archives Council (MLA, 2008a, 2008b, 2011; MLA and Local Government Group, 2011); Arts Council England (Smithies, 2011; Davey, 2013; the Reading Agency, 2013), the Scottish Executive (Scottish Executive and Scottish Library and Information Council, 2007), the Scottish Library and Information Council (SLIC, 2011a; 2011b, 2014, 2015), the Welsh Government and CyMAL (Welsh Assembly Government, 2011b; Welsh Assembly Government and CyMAL, 2011; Welsh Government, 2014b), the Department of Culture, Arts and Leisure, Northern Ireland (DCAL, 2007) and Libraries NI (2010, 2011, 2014).

In England, as the Carnegie report stated, strategy and policy-making for public libraries is 'essentially a local service' (Macdonald, 2012, p. 49), but a number of bodies exercise strategic influence at a national level. At the centre of activity are the UK Government's Department for Culture, Media & Sport (DCMS), and the principal agencies for steering the development of public libraries, specifically the Museums, Libraries and Archives Council (MLA) until 2011, and currently the Arts Council England (ACE). From around 2000 onwards, these agencies have played an active role at national level in identifying key issues in the public library sector and in driving initiatives to ensure that the public library system remains up-to-date and relevant to the 21st century.

The strategic documents cited above for English public libraries convey considerable debate, some of it contradictory, about the nature of the 'core offer' in modern libraries. The Modernisation Review issued by the DCMS in 2010 makes specific proposals for inclusion in libraries' planning (DCMS, 2010, pp. 13–14), although ACE in 2011 stresses that the question of libraries' core role is still contested (Smithies, 2011, p. 15). By contrast, Wales and Scotland have identified core provision for public libraries, and have specified in some detail the definitions for each element (Scottish Executive and Scottish Library and Information Council, 2007; SLIC, 2015; Welsh Assembly Government and CyMAL, 2011; Welsh Government, 2014c). Northern Ireland also applies a set of clearly defined standards (DCAL, 2007). Whatever might be the outcome of ongoing debates about core provision, the broad role for today's public libraries can be discerned from these policies. The different versions in each of the four jurisdictions do share significant common elements. They clearly reflect contemporary political and social issues, including those expressed in the government digital policies described earlier in this chapter. They embrace information provision; services; location and premises; social inclusion; community engagement and well-being; promotion of reading skills; digital inclusion and participation; digital literacy; support for informal and formal learning; and use of new technology to provide innovative services and financial efficiencies. Scotland, Wales and Northern Ireland also include an explicit duty to provide and promote materials relating to their countries' culture and language.

It is notable that until 2010, a consistent element in England's deliberations was the view that books were still core to public library provision, despite the onward march of digital provision. The DCMS' Modernisation Review (DCMS, 2010), however, expressed the role of digital provision more fully and clearly. With the *Library 21* research commissioned by ACE (Reading Agency, 2013), the same organisation's research programme *Envisioning the library of the future* (ACE, 2013), and the *Independent Library Report* (DCMS, 2014a), the provision of blended digital and physical information resources had become a high priority, and is fully endorsed in the DCMS' report for the same year (DCMS, 2014b). This unease about the changing role of the book in digital times was not evident in the planning of Scotland, Wales and Northern Ireland. In these jurisdictions, the emphasis has consistently been on the provision of information resources in all relevant formats and media, including digital content.

At the level of strategic statements about digital provision, the priorities for English and Scottish public libraries seem to be on supporting digital access, inclusion and literacy and on providing digital content that has been sourced largely from external suppliers (whether commercially funded or publicly available from non-commercial providers), rather than on digital content creation and digitisation within their own remit. For England there are a few references to libraries creating their own digital materials (DCMS, 2003, pp. 36–37 and 2010, p. 11; Reading Agency, 2013, p. 9), but this is not emphasised in the overall strategic thinking. The Welsh position is markedly different. Digitisation is specifically included in their policies, and is given significant prominence (Welsh Assembly Government, 2011b; Welsh Government, 2014b, 2014c). Welsh libraries are expected to 'reflect changing forms of publication' and to achieve access to the resources of all Welsh libraries by (amongst other actions) cooperating 'to create new bilingual digital content about Wales and its people' (Welsh Assembly Government, 2011b, pp. 9, 21). The most recent government review of Welsh public libraries focuses primarily on the very real challenges of the financial climate and consequent remodelling of the public library system, not least in the light of current technology. Nonetheless, it also reaffirms the role of digital materials, and in relation to this, the value of libraries being able to 'create and publish [their] own information resources' (Welsh Government, 2014b, p. 39). In Northern Ireland, digitisation's importance has become more explicit in most recent years. The plans issued by Libraries NI refer to digitisation as an active element of public libraries' work (Libraries NI, 2010, 2014), and this is supported by the creation of a digitisation strategy issued in 2015 (Libraries NI, 2014, 2015).

2.6 MUSEUMS

The strategic oversight of the museum sector is the responsibility of the DCMS for England and of the devolved governments for Scotland, Wales and Northern Ireland.

As the Museums Association reports, 'there have been several attempts to produce a national museum strategy for England, but these have been unsuccessful' (Museums Association, c2015). In practice, government control in England operates at two levels. The DCMS oversees policy for the national museums, which are responsible for their own strategic planning (DCMS, 2015). Non-national and regional museums have been steered on behalf of DCMS firstly by the MLA until its demise in 2011, and since then by ACE. There has been a major, long-running programme for museums development in the form of the 'Renaissance in the Regions' programme. This was initiated in 2002 to create a national framework for museums and galleries; to drive long-term sustainability; to ensure a valued social purpose for museums; and to develop a well-managed national collection. In addressing these strategic goals, the programme's main aims were for museums to be an important resource for learning and education, enabling access and social inclusion; functioning as focal points in local communities; contributing to economic regeneration and collecting; and caring for and interpreting the material culture of UK. A full-scale review of the programme in 2009 established that there had been a strong start in many aspects of this ambitious programme, but it also identified a range of important issues requiring attention for the sector to move forward as desired (*Renaissance in the Regions...*, 2009). At this stage, there was little specific reference to the digital agenda, although its role can be inferred in various sections. This emerged more clearly in the MLA's subsequent *Museum Action Plan*, devised to address the findings of the above review. This Plan's aims specifically included investing in the digital future, albeit presenting a general and overarching statement of the role of digital technology in museums (MLA, 2009). When ACE took on the leadership of the sector, it issued an initial strategy for museums and libraries (ACE, 2011), setting out short-term directions for the sectors under ACE's oversight. It also gave first indications of the likely longer term strategy: excellence; inspiring people to greater use of the sector's assets; encouraging use by children and young people; sustainability; resilience; and skilled leaders and staff. The intention is to develop the strategy further in consultation with relevant stakeholders, and ultimately to establish one integrated strategy for arts, museums and libraries. The interim strategy states clearly that 'the digital agenda is a challenge and opportunity across the arts, museums and libraries' (ACE, 2011, p. 12). While digital content creation and digitisation do not feature explicitly in this document, they are an essential means of delivering the specific requirements identified, notably the calls for museums and libraries to innovate in the use of digital media, and to meet communities' need for online as well as physical access to collections. The political support for such development was reinforced in a speech by Sir Peter Bazalgette, Chair of ACE, where he stressed that 'the digital revolution is giving museums a great opportunity to reach more people in more places in more ways...museums have been at the forefront of this, in thinking how they can make their collections available on the web, how they can interact with the public digitally and how they can involve people in curating content' (Bazalgette, 2013).

In Scotland, the museum sector is overseen by Museums Galleries Scotland, formerly the Scottish Museums Council. Their planning shows that there has been strategic recognition of the role of digital content and digitisation since at least 2000, when work began on a national ICT strategy (Scottish Museums Council, 2004). This was created to promote and support the use of ICT in Scotland's museums, building on a strong existing base of successful initiatives in Scotland, and recognising that the digital agenda would become ever more important in the following years. Its aims included increasing the level of users' access by online means; mainstreaming technology in Scottish museums; and driving up the quality of electronic output. Although this strategy addressed the typically wide range of ICT issues of the time, its recognition of digital content creation as a necessary part of museum activity was clearly and explicitly stated throughout the document. The evident commitment was reinforced in 2005, with the Council's complementary publication *Museums, Galleries and Digitisation* (Scottish Museums Council, 2005). This was a clearly articulated guide to best practice in policy-making for digitisation in museums as well as providing recommendations on measuring its impact. It was followed in turn by the *Digital Content Action Framework* which itemised the various stages of planning and implementing a digitisation project (Museums Galleries Scotland, 2008).

This body of work was well established by the time the most recent, fully revised strategy for the Scottish museums and galleries sector was drawn up. *Going Further*, published in 2012, carries a very dynamic vision for the next ten years in the country's museums and galleries. This is to be achieved by maximising the potential of Scotland's collections and culture; strengthening connections between museums, people and places; and developing a global perspective, all of which are to be supported by an empowered, diverse workforce, sustainability and collaboration (Museums Galleries Scotland, 2012). Within these broad aims, it is made clear that virtual collections and the related use of technology will be part of that future. In line with this, the supporting delivery plan, issued the following year, includes specific actions to develop online collections and digital resources (Museums Galleries Scotland, 2013).

The museums sector in Wales comes under CyMAL (Museums Archives and Libraries Wales), which brings together its constituents very actively to support development and collaboration within and between the sectors. In 2010, CyMAL issued the first national museum strategy for Wales, covering the period to 2015 (CyMAL, 2010). This embraces the many aspects of museum work, building its action plan around three main strands: museums for everyone; a collection for the nation; and working effectively. Each of these is supported by a range of actions that address the multifaceted nature of museum work, and of the developments to be addressed in the planning period. The strategic emphases include the development of a distributed national collection; accessibility to all; supporting both formal and informal learning; responding to the needs of children and young people; and public participation in the creation and development of the country's museums. The recognition of the digital environment within this is explicit: 'in the 21st century, museum objects can be material or virtual' (CyMAL, 2010, p. 22). Access to websites and virtual tours feature in the formal, national evaluation carried out a year after the strategy was launched (CyMAL, 2013). Most noticeably, the strategy articulates clearly the role of digital materials in the development of the *People's Collection Wales*, which is presented as 'a strategic approach to providing online digital content' (CyMAL, 2010, p. 21), to be backed up by robust research and public participation, and to be supported by the exploitation of advanced technology to enable personalised use of the content. While digital content and digitisation do not feature explicitly in these sections, they number amongst the requirements to deliver the strategic directions outlined in the document.

In Northern Ireland, the Department of Culture, Arts and Leisure is responsible for museums. Its overall vision and strategic aims for the sector are outlined in the Northern Ireland Museums Policy (DCAL, 2010). This seeks to achieve a coordinated, sustainable museum sector that develops, preserves and interprets its collections; delivers quality services to inspire and educate; and draws on the region's strengths and diversity to support economic, social and cultural development in Northern Ireland. Its strategic priorities focus on developing audiences; supporting education and learning (whether formal or lifelong); collections development, care, management and use; and provision of the infrastructure, investment and resources to support these aims. It places considerable emphasis on aspects of provision such as access, user needs and public engagement with collections. It is in this context that the policy notes the importance of including digital technologies 'as a means of capturing and disseminating…information' (DCAL, 2010, p. 15). This in turn is part of a clearly stated objective to maximise the use of digital and emerging technologies to enhance museums and their services (DCAL, 2010, p. 17). As with similar strategies in some other regions, it can be seen that digital content creation and digitisation are both relevant means of achieving these ends.

2.7 ARCHIVES

2.7.1 The National Archives

At national level, the archives sector in England and Wales is led by The National Archives (TNA). This organisation came into existence in 2003 when the Public Record Office and the Historical Manuscripts Commission merged. The Office for Public Service Information subsequently amalgamated with it in 2006.

In 2006, TNA issued *A New Vision for The National Archives 2006—11*, stating the organisation's purpose as being to lead and transform information management; to guarantee the survival of today's information for tomorrow and to bring history to life for everyone (TNA, 2008a, p. 11). An updated document, *Living Information: the Vision of The National Archives* (TNA, 2007a), elaborated on these principles. It presented a prominent and wide-ranging role for TNA as a policy leader, advisor and facilitator in information management for both the government and the wider public sector. This included their 'commitment to providing people worldwide with access to our records, and to helping everyone use them to excite and enrich their lives' (TNA, 2007a, p. 5). Significantly, in fulfilling these aims TNA expected over time to become a predominantly online service delivery organisation.

The online strategy issued in 2008 addressed the requirements of these aspects of TNA's Vision, and it covered the many facets of digital curation. There was a clear emphasis on the management of the full range of the organisation's website, including the presentation and retrieval of digitised content. The Strategic Plan from the same period (TNA, 2007b) presented a strong, proactive five-year strategy to 2012 and stressed the major environmental changes affecting archives management and delivery of services, many of which were driven by or were a consequence of the new technologies. Issues included the survival of digital information, and online provision and digitisation, with references throughout the document to the nature of work required in these fields to meet current and future needs. These elements were carried forward consistently in subsequent planning (TNA, 2008b, 2009c, 2010a, 2010b, 2011) and in the organisation's *Digital Strategy* (TNA, 2012b).

In 2009 the UK Government published *Archives for the 21st Century* (HM Government, 2009). This was the first statement of national policy for the archive sector as a whole since 1999, and an essential renewal of vision and strategy, given the transformational changes in the wider environment since the last such document. It focused on the power of archives in creating national, community and individual identity; their importance in formal education as well as academic and personal research; and their role in sourcing evidence to demonstrate integrity and judgement of public and private decisions and actions. This detailed document presented the vision of realising the true potential of publicly funded archives, identified the obstacles to achieving this end, proposed how to achieve it and outlined actions to take it forward. The digital context was strongly emphasised throughout, with implicit and explicit references to the role of digitisation. It acknowledged 'a growing expectation in society that all information resources should be available online, 24 hours a day' (HM Government, 2009, p. 6), and that 'work to digitise collections is as yet mostly small-scale and piecemeal' (HM Government, 2009, p. 5). It included in its key recommendations the commitment to providing digitised archive content (HM Government, 2009, p. 16). This was reaffirmed in the subsequent action plan issued by The National Archives and the Museums, Libraries and Archives Council in 2010 (TNA and MLA, 2010) and in the update to this plan published in 2012 (TNA, 2012a). Significantly, all these documents encouraged digitisation projects throughout the sector.

There is no separate public record office holding government records for Wales. Under the terms of a formal concordat (Welsh Government, 2011), TNA performs this role, while the archives held by local authorities, universities and specialist centres operate, as in England, under the direction of their relevant parent bodies. In recent years, there has been growing attention to the potential for creating a separate National Archive for Wales, and the NLW is a leading organisation in promoting this debate (NLW, 2014a, p. 10).

As a key document for the future of national services, a separate version of *Archives for the 21st Century* was issued to reflect the variations required in provision for Wales (Welsh Assembly Government, 2009). The core vision and issues are the same in the versions for both England and Wales, with the latter also covering areas specific to Welsh culture and language. It is notable, however, that the relevance of digital developments is stressed in the Introduction by the Welsh Minister for Heritage (Welsh Assembly Government, 2009, pp. 1−2).

2.7.2 National Records of Scotland

The National Archives of Scotland (NAS) merged in 2011 with the General Register Office for Scotland (GROS) to form the National Records of Scotland (NRS). The Corporate Plans of the NAS set out annual aims and objectives for each of the organisation's divisions (NAS, 2003–13). From the earliest years of the last decade, digital content, both born-digital and digitised from analogue materials, becomes increasingly prominent. A very strong focus on major digitisation emerges, reflected in groundbreaking initiatives such as *ScotlandsPeople*, *ScotlandsPlaces*, and the *Scottish Archive Network (SCAN)*, as well as online access within these to a significant range of key resources illustrating other aspects of Scottish society and heritage. The centrality of such initiatives is reinforced from 2008 onwards when the plans articulate clearly their alignment with the strategic issues at the heart of the Scottish Executive's thinking, including the social, educational and economic value of the nation's archives, and their central role in defining and illustrating Scottish identity. The NRS' *Strategy, 2012–22,* contrasts with the most recent TNA documents, in that digital provision is embedded throughout the paper, serving as an implicit rather than explicit element of the organisation's strategic directions (NRS, 2012). The continuing commitment to digital content and digitisation is, however, clear from the NRS' *Business Plan* (NRS, 2011) and the organisation's *Records Strategies* (NAS, n.d.).

2.7.3 Public Record Office of Northern Ireland

By the early years of the last decade, in line with the other UK archive organisations, the Public Record Office of Northern Ireland (PRONI) was also engaging with 'the need to embrace and make effective use of new technology for the benefit of customers' (PRONI, 2003, p. 3). This plan recognised the value of digitisation as a means of making available important heritage items, and material in most demand from customers. It committed to a programme of digitised genealogical and other information resources for online access. The new *Archives Policy for Northern Ireland*, initiated in the following year, reaffirmed the strategic importance of a properly planned programme based on consultations with customers (DCAL, 2004, p. 13). As the organisation's annual reports demonstrate, this commitment led to regular digitisation initiatives in the following years, some in collaboration with other regional organisations that continue to the present day (PRONI, 2006–14).

2.8 INDEPENDENT ORGANISATIONS

2.8.1 Wellcome Trust

The Wellcome Trust is an independent organisation established in 1936 to foster and promote research for the improvement of human and animal health. Its activities are characterised by its consistent focus on large-scale and global issues and correspondingly ambitious initiatives to further research that is relevant to these. As an extremely well-funded and independent Trust, it aims to tackle major issues responsibly and flexibly, to the extent that it will take a long-term view and accept funding risks to achieve its goals (Wellcome Trust, 2005, p. 4).

Planning for the Future, the title of the Trust's *Strategic Plan* for 2000–2005, reflects this overall approach. At that period, the emphasis was on enhanced clinical research; public engagement activities; translation of research findings to applications for the improvement of health; and international research. Major initiatives at the time included their work on the Human Genome Project and the Beowulf Genomics programme. Research into the history of medicine was also an important part of this plan, as an essential element in understanding and advancing current health issues. Consequently, the Trust supported significant programmes for building databases as well as cataloguing, conserving and preserving information materials, but it was not until the next *Strategic Plan* that digitisation emerged specifically (Wellcome Trust, 2005). From this stage onwards there were explicit and repeated commitments to digitisation in both the organisation's plans and its annual reports (Wellcome Trust, 2003–2013, 2014), along with a clear understanding of the support needed to effect the work. In the *Annual Report* for 2007, it was announced that the Wellcome Library, already established as a world-leading research facility for the history of medicine, would embark on a major project to digitise its holdings systematically to increase global access to its collections (Wellcome Trust, 2003–2013, see 2007, p. 8). This first large-scale initiative was completed in four years (Wellcome Trust, 2012, p. 12).

By this time, the Trust had developed a new, ten-year *Strategic Plan* (Wellcome Trust, 2010), expressly to establish a truly long-term view of global health issues and to initiate large research plans that would produce 'ambitious approaches to make progress' (Wellcome Trust, 2010, p. 2). The Plan identified five major challenges: achieving the health benefits of genetics and genomics; understanding the brain; combating infectious disease; understanding development, ageing and chronic disease; and connecting the environment, nutrition and health. To address these, it also committed to supporting outstanding researchers, and this includes maximising the global reach of the Wellcome Library through digitisation. The level of commitment is demonstrated by the inclusion of a specific key indicator to measure the organisation's contribution to the creation, development and maintenance of major research resources (Wellcome Trust, 2010, p. 18). Within this context, the Wellcome Library's digitisation initiatives have been developed to a new level. Following the major digitisation programme launched in 2007, the Library is currently aiming to digitise as much as possible of its holdings (Wellcome Library, 2015b).

2.9 STRATEGIC CONTEXT IN SUMMARY

At first sight, the details of strategic thinking about digitisation, as summarised above, might appear to endorse fully the views reported in Chapter 1 that publicly and philanthropically funded digitisation has been disjointed and carried out in silos over the last 20 years. Certainly, the complex environment does contribute to such an outcome. It has functioned at three different levels: governments, government agencies, and individual institutions. These have resulted in strategies which do not interrelate consistently and which vary considerably in their positions about digital content creation and digitisation.

At the government level, they do present all common strands regarding the impact of the digital environment, the development and exploitation of digital economies and the central role of the ever-increasing digital information base. There have been marked inconsistencies amongst the different administrations, however, in relation to digitisation. In the case of the UK government, there is a timeline running through the last decade during which recognition of the value of digitisation becomes more explicit, culminating in 2009 with *Digital Britain's* praise for the work of heritage and knowledge organisations throughout the country. This support, however, is of a rather passive nature, and does not reappear in subsequent UK strategies. By contrast, the devolved administrations have stressed the importance of digital content and digitisation in successive strategies and related business plans. The Welsh strategies give digitisation a particularly prominent role, presenting it as firmly embedded in the country's methods for advancing its desired social and educational development.

If the Westminster government's strategic position is noncommittal, this is somewhat mitigated by its agencies that steer or influence the strategic development of libraries, museums, galleries and archives. In the higher education sector, Jisc has carried active responsibility for promoting and implementing digitisation relevant to the sector, and the AHRB/C has driven forward developments within the discipline of Digital Humanities. In the case of both these organisations, their most recent strategies carry shifts in emphasis that move beyond digitisation work, thereby repositioning it as an implicit means of delivering their strategies, rather than a strategic end in itself.

Digitisation, therefore, has been an area of major strategic interest in the period from 2000-2010 approximately, but by the beginning of the current decade it had lost much of its earlier strategic support, despite the continuing necessity to bring analogue content into the digital environment to deliver the digital provision expected by both government and the wider population. It is clear, however, that whatever the variant levels of government support, the heritage and knowledge organisations in the UK have not lost sight of digitisation's role, and remain committed to its advancement. All the national libraries have taken a long-term view and accordingly have sustained digitisation in their strategic planning both past and present. Similarly, the museums sector has long positioned digitisation within its mainstream planning. Currently, this is particularly marked in both Wales and Northern Ireland, while the strategies of MLA and ACE since 2011 have brought England more closely into line with the devolved nations. In the archives sector, digitisation has been at the heart of TNA's strategies to widen access since the last decade, and NAS/NRS have followed a systematic digitisation programme over much the same period. Only in the public library sector in England and Scotland is there a variant strategic approach, with the delivery of digital content being focussed principally on accessing material provided by commercial suppliers or made available by other organisations, rather than prioritising digitisation by the libraries themselves. It is important to note, however, that the Welsh and Northern Irish strategies do actively encourage digitisation within the deliverables of their libraries.

Overall, therefore, the majority of publicly funded library, museum and archive services do recognise a duty to digitise. It is fair to say, however, that the prioritisation of such activities has been constrained in many of the strategies by the wide range of strategic obligations on the services, and the limited financial resources available for the costly business of digitising. In the independent, philanthropic sector, such issues can also arise, but, without the same pressure to respond to political circumstances, choices for prioritisation can result in very different outcomes. The Wellcome Trust's approach is an outstanding example of this. Of all the strategies considered in this chapter, it is the most ambitious and the most confident in terms of its visionary commitment to large-scale digitisation. This is a direct result of the organisation's independence, and its strong financial position.

2.10 COLLABORATION

The organisations covered in this chapter vary in their strategic positions concerning digital content creation and digitisation in particular. At the overall strategic level, however, they are all markedly consistent in their commitment to collaboration, and several explicitly include digitisation within their plans for shared developments. The terminology does vary, with 'partnership', a frequent choice in place of 'collaboration', while 'cooperation', 'coordination', 'sharing' and in at least one instance 'co-working' also feature. The differences in nuance reflect in turn differences in emphasis between the strategies but not in their fundamental principles, and for the purposes of this book, 'collaboration' is used when discussing this aspect of the organisations' activities.

Within this overall consistency, there are variations in the range and depth of the commitments to collaboration. This can be seen in the most commonly cited drivers for collaboration at the overall strategic level, as expressed in the documents. Unsurprisingly, for many it is a means of seeking value for money, financial efficiency, savings, and/or additional funding. The motivations do, however, extend well beyond this most obvious of reasons. There is also a strong emphasis on the opportunities to achieve more by working with others than can be achieved by working alone, and this is specified by the library and archives sectors in particular. The public libraries and museums are prominent among those targeting improved public services through collaboration, and in sharing collections and infrastructure. Innovation also features strongly, especially in the plans of the museums and archives. Beyond these, several additional reasons feature over time in the thinking of individual organisations. For example, HEFCE places value on collaboration to produce content of all sorts (Becta, 2008; HEFCE; Online Learning Task Force, 2011). AHRC sees collaboration as a means of increasing impact and public engagement (AHRC, 2014). TNA emphasises its potential to improve stewardship and to promote a sense of identity and place in a community (TNA and MLA, 2010). In all these cases, the different individual factors reflect directly priorities in the overall strategic perspective of their organisations. In general terms, there is no one pattern of strategic drivers for collaboration, but rather many permutations, with powerful potential and complex inter-relationships. TNA explores this environment in some detail from the perspective of archive services, but many of its observations are equally relevant to the other strategies and organisations discussed in this chapter (TNA and MLA, 2010, pp. 16, 18).

There are also variations to be found in the range and background of partners that the institutions seek for their collaborative ventures. In a well-established and familiar formula, most subscribe, in one way or another, to the broad combination of working with others in their own field and in other public sectors, as well as looking to the business, private and charity sectors. Most also seek collaboration at national and international levels. While some organisations choose not to elaborate further, some illustrate in detail an extensive range of actual or potential partners. Jisc, for example, stresses its strong relationships with UK research and educational in-stitutions, funders, government agencies, enterprise and industry sectors, and relevant organisations in Europe and around the world (Jisc, 2012). The Funding Councils, AHRC, British Library and TNA present equivalent groups of collaborators. By contrast, some organisations set a tighter focus for their partnerships, albeit within the wider context already described. NLS, NLW and NRS, for instance, focus particularly strongly on their own nations and their own sector as well as on other heritage organisations and the academic community. The museums typically concentrate strongly on other museums, the wider heritage and tourism sector, and the players in the creative economy.

Within this overall context, many of the organisations reinforce their commitment to collaboration by specifying the partners and programmes with which they are or will be working in the period of their strategies. In the cross-sectoral arena, a particularly strong partnership, initiated at gov-ernment level, emerges from the Welsh organisations, where NLW, the public libraries and the museums all prioritise their involvement in the ongoing and very active all-Wales collaboration that has already been observed earlier in this chapter. Several of the strategies are also specific in relation to digitisation in particular. The British Library identifies its cross-sectoral partners as including 'publishers, external funding bodies, other sponsors and donors, technology providers, user communities and rights holders' (British Library, 2008). In the course of several strategies, the Li-brary reports significant, ongoing involvement in all levels of work: mass digitisation with Microsoft and Google; large-scale projects with Jisc and, increasingly, smaller scale and subject-focused initiatives with research organisations and individuals. NLS identifies the need to work with Scottish partners to develop a national approach to digitisation (NLS, 2005) and its strategies and plans include details of shared initiatives both within and beyond Scotland. NLW cites several major all-Wales projects as well as its Digital Collections Research Programme (NLW, 2014b, 2015). TNA reports a range of major UK partners (TNA, 2012a) in their *Annual Reports*, while the NAS and NRS describe in detail a series of shared digitisation programmes with other, mostly Scottish, organisations (NAS, 1999–2007; NRS, 2008–14). The Wellcome Trust explicitly commits to including others' content in its digitisation programmes where it is complementary to its holdings (Wellcome Library, 2015a).

Collaboration at the broadest levels, as described at the beginning of this section, represents a creditably ambitious approach, with many successful outcomes. In practice, all the organisations covered here have to make choices and to set priorities in collaborative work as in all other activities. In most cases, it is implicit that such decisions are made by each to remain in keeping with its overall strategy. As noted in the previous paragraph, there are examples of initiatives where individual organisations agree to work beyond their immediate strategic interests to meet the needs of other partners, but such instances remain within the extended boundaries of their strategic context. By contrast, however, there are a few instances where the need for a tight strategic focus is made quite explicit, such as in the NLW's long-term vision, where future activities are to be concentrated 'on outcomes contributing centrally and substantially to the Library's mission, core functions and strategy' (NLW, 2010, p. 29). The Wellcome Trust is equally clear when it states that its emphasis in collaborative work will be on leadership and brokering the right partnerships to further its mission (Wellcome Trust, 2010, p. 17).

The fact that collaborative work must meet the strategic direction of each partner represents both strengths and weaknesses in such activity. There are major advances in areas of common interest that can be and are taken forward only through collaboration, and as noted there are levels of flexibility brought to such work by partners to avoid unnecessary obstacles. There are also several examples of partnerships formalised by means of Memoranda of Understanding (MoUs), (British Library, 2010; Jisc, 2005) or joint statements (AHRC, 2010). These allow the work to be positioned constructively within the relevant strategies and thereby they strengthen the actual initiatives taken.

Ultimately, however, there will be limits to every organisation's options, and there will be initiatives that cannot go forward because they lack the collaborative commitment that they require. Collaboration is a finely balanced and delicate environment, and one of Jisc's descriptions of its own partnerships could well be applied to many other organisations: '[collaboration means working] with a coherent and complex set of regional, national and international partnerships' (Jisc, c2007).

REFERENCES

Arts and Humanities Research Council. (c2007) *AHRC Vision and Strategy 2007-2012.* Bristol: AHRC. [Online] Available from: http://www.ahrc.ac.uk/News-and-Events/Publications/Documents/Vision-and-Strategy 2007-2012.pdf.

Arts and Humanities Research Council. (2010) *Delivery Plan 2011-2015.* Swindon: AHRC. [Online] Available from: http://www.ahrc.ac.uk/News-and-Events/Publications/Documents/Delivery-Plan-2011-2015.pdf.

Arts and Humanities Research Council. (2013) *The Human World: the Arts and Humanities in our time. AHRC Strategy 2013-2018.* Swindon: AHRC. [Online] Available from: http://www.ahrc.ac.uk/News-and-Events/News/Documents/AHRC-Strategy-2013-18.pdf.

Arts and Humanities Research Council. (2014) *Delivery Plan 2015-2016.* Swindon: AHRC. [Online] Available from: http://www.ahrc.ac.uk/News-and-Events/Publications/Documents/AHRC%20Delivery%20Plan%202015-16%20(A).pdf.

Arts Council England. (2011) *Culture, Knowledge and Understanding: great museums and libraries for everyone.* [Online] Available from: http://www.artscouncil.org.uk/media/uploads/pdf/culture_knowledge_ and_understanding_final010312.pdf.

Arts Council England. (2013) *Envisioning the Library of the Future.* [Online] Available from: http://www.artscouncil.org.uk/what-we-do/supporting-libraries/other-links/library-of-the-future/.

Bazalgette, Sir P. (2013) *Speech*, Museums Association Conference, Liverpool, 12 November 2013. [Online] Available from: http://www.artscouncil.org.uk/media/uploads/pdf/Peter_ Bazalgette_speech_Museums_Association_conference.pdf.

Becta. (2008) *Harnessing technology: next generation learning 2008-2014.* [Online] Available from: http://dera.ioe.http://webarchive.nationalarchives.gov.uk/20060315075935/http://dfes.gov.uk/publications/docs/e-strategy.pdfac.uk/8287/1/download.cfm%3FresID%3D37348.

Brindley, L. (2000) Taking the British Library forward in the twenty-first century. *D-Lib Magazine.* Vol. 6, Issue 1. [Online] Available from: http://www.dlib.org/dlib/november00/brindley/11brindley.html.

British Library. (2001) *Twenty-eighth Annual Report and Accounts 2000-01.* London: British Library Board. [Online] Available from: http://www.bl.uk/aboutus/annrep/0001/28thannualreport.pdf.

British Library. (2005) *Redefining the Library: the British Library's Strategy 2005-2008.* London: British Library. [Online] Available from: http://www.bl.uk/aboutus/foi/pubsch/strategy_0508.pdf.

British Library. (2008) *Digitisation Strategy 2008-2011.* [Online] Available from: http://www.bl.uk/aboutus/stratpolprog/digi/digitisation/digistrategy/.

British Library, (2010) *2020 Vision.* [Online] Available from: http://www.bl.uk/aboutus/stratpolprog/2020vision/.

British Library. (2011) *Growing knowledge: the British Library's strategy*. London: British Library. [Online] Available from: http://www.bl.uk/aboutus/stratpolprog/strategy1115/strategy1115.pdf.

British Library. (c2015) *Living Knowledge: the British Library 2015-2023*. London: British Library. [Online] Available from: http://www.bl.uk/britishlibrary/~/media/bl/global/projects/living-knowledge/documents/living-knowledge-the-british-library-2015-2023.pdf.

British Library. (n.d.) *Digitisation*. [Online] Available from: http://www.bl.uk/aboutus/stratpolprog/digi/digitisation/.

Cabinet Office: Prime Minister's Strategy Unit and Department of Trade and Industry. (2005) *Connecting the UK: the Digital Strategy*. London: Strategy Unit. [Online] Available from: http://www.bis.gov.uk/files/file13434.pdf.

Carpenter, L., Shaw, S. and Prescott, A. (Eds.) (1998) *Towards the Digital Library: the British Library's Initiatives for Access Programme*. London: British Library.

CyMAL Museums Archives and Libraries Wales. (2010) *A Museums Strategy for Wales*. [Online] Available from: http://wales.gov.uk/docs/drah/publications/100615museumstrategyen.pdf.

CyMAL Museums Archives and Libraries Wales. (2013) *Spotlight on Museums 2011*. [Online] Available from: http://wales.gov.uk/docs/drah/publications/130503spotlight2011en.pdf.

Davey, A. (2013) *The Library of the Future: a response to 'Envisioning the Library of the Future'*. London: ACE. [Online] Available from: http://www.artscouncil.org.uk/media/uploads/pdf/The_library_of_the_future_May_2013.pdf.

Denbo, S., Haskins, H. and Robey, D. (2008) *Sustainability of Digital Outputs from AHRC Resource Enhancement Projects*. [Online] Available from: http://www.ahrcict.rdg.ac.uk/activities/review/sustainability08.pdf.

Department for Culture, Media & Sport. (2003) *Framework for the Future: libraries, learning and information in the next decade*. London: DCMS. [Online] Available from: http://webarchive.nationalarchives.gov.uk/+/http:/www.culture.gov.uk/reference_library/publications/4505.aspx.

Department for Culture, Media & Sport. (2010) *The Modernisation Review of Public Libraries: a policy statement*. London: DCMS. [Online] Available from: http://zdruzenje-knjiznic.si/media/website/modernizacija-splosnih-knjiznic-velika-britanija/UK-public-lib-policy-statement_2010.pdf.

Department for Culture, Media & Sport. (2012) *Science and Heritage: Response to the House of Lords Select Committee on Science and Technology*. London: Stationery Office. Cm 8384. [Online] Available from: https://www.gov.uk/government/publications/science-and-heritage-response-to-the-house-of-lords-select-committee-on-science-and-technology-2012.

Department for Culture, Media & Sport. (2013a) *Connectivity, Content and Consumers: Britain's Digital Platform for Growth*. London: DCMS. [Online] Available from: https://www.gov.uk/government/uploads/system/uploads/attachment_data/file/206944/13-901-information-economy-strategy.pdf.

Department for Culture, Media & Sport. (2013b) *Report under the Public Libraries and Museums Act 1964 for 2012/2013*. London: DCMS. [Online] Available at https://www.gov.uk/government/publications/report-under-the-public-libraries-and-museums-act-1964-for-201213.

Department for Culture, Media & Sport. (2014a) *Independent Library Report for England*. [Online] Available from: https://www.gov.uk/government/uploads/system/uploads/attachment_data/file/388989/Independent_Library_Report-_18_December.pdf. London: DCMS.

Department for Culture, Media & Sport. (2014b) *Report under the Public Libraries and Museums Act 1964 for 2014*. London: DCMS. [Online] Available from: https://www.gov.uk/government/uploads/system/uploads/attachment_data/file/387946/Annual_Library_Report_-_16_Dec.pdf.

Department for Culture, Media & Sport. (2015) *2010 to 2015 government policy: museums and galleries*. [Online] Available from: https://www.gov.uk/government/publications/2010-to-2015-government-policy-museums-and-galleries/2010-to-2015-government-policy-museums-and-galleries.

Department for Culture, Media & Sport and Department for Business, Innovation & Skills. (2009) *Digital Britain: Final Report*. London: Stationery Office. Cm 7650. [Online] Available from: https://www.gov.uk/government/uploads/system/uploads/attachment_data/file/228844/7650.pdf.

Department for Education and Skills. (2005) *Harnessing technology: transforming learning and children's services*. Nottingham: DfES. [Online] Available from: http://webarchive.nationalarchives.gov.uk/20130401151715/https://www.education.gov.uk/publications/eOrderingDownload/1296-2005PDF-EN-01.pdf.

Department for Employment and Learning, Northern Ireland. (c2012) *Graduating to success: a higher education strategy for Northern Ireland*. [Online] Available from: http://www.delni.gov.uk/graduating-to-success-he-strategy-for-ni.pdf.

Department of Culture, Arts and Leisure, Northern Ireland. (2004) *Archives Policy for Northern Ireland Consultation Document and Marketing Strategy for the Public Record Office of Northern Ireland*. [Online] Available from: http://www.dcalni.gov.uk/archive_policy_short-2.pdf.

Department of Culture, Arts and Leisure, Northern Ireland. (2007) *Delivering Tomorrow's Libraries: principles and priorities for the development of public libraries in Northern Ireland*. Belfast: DCAL. [Online] Available from: http://www.dcalni.gov.uk/final__delivering_tomorrow_s_libraries__document_-_july_2007_-_1mb_document_for_website.pdf.

Department of Culture, Arts and Libraries, Northern Ireland. (2010) *Northern Ireland Museums Policy*. [Online] Available from: http://www.dcalni.gov.uk/pdf_version_of_final_museums_policy.pdf.

Department of Trade and Industry. (1998) *Our Competitive Future: Building the Knowledge Driven Economy*. London: DTI. Cm 4176. [Online] Available from: http://webarchive.nationalarchives.gov.uk/+/http://www.dti.gov.uk/comp/competitive/wh_int1.htm.

Digital Northern Ireland Advisory Board. (2010) *Digital NI 2020: from connectivity to interactivity*. [Online] Available from: http://www.dni2020.com/home.

Dunning, A. (2010) Digitizing the past: next steps for public sector digitization. In Woodward, H. and Estelle, L (eds) *Digital Information: order or anarchy?* London: Facet: 117–131

Glenaffric. (2008) *Review of the 2005 HEFCE Strategy for e-Learning: a report to HEFCE*. Glenaffric. [Online] Available from: http://www.hefce.ac.uk/media/hefce/content/pubs/indirreports/2008/missing/Review%20of%20the%202005%20HEFCE%20Strategy%20for%20e-Learning.pdf.

Green, A. (2002) Digital Library, Open Library: developments in the National Library of Wales. *Alexandria*. Vol. 14, No. 3: 161–70. [Online] Available from: http://www.llgc.org.uk/fileadmin/documents/pdf/alexandria_S.pdf.

Higher Education Funding Council for England. (2009) *Enhancing Learning and Teaching through the use of Technology: a revised approach to HEFCE's strategy for e-learning*. HEFCE. HEFCE 2009/12. [Online] Available from: http://www.hefce.ac.uk/media/hefce1/pubs/hefce/2009/0912/09_12.pdf.

Higher Education Funding Council for England. (2011) *Opportunity, choice and excellence in higher education*. HEFCE. HEFCE 2011/22. [Online] Available from: http://www.hefce.ac.uk/media/hefce/content/about/howweoperate/corporateplanning/strategystatement/HEFCEstrategystatement.pdf.

Higher Education Funding Council for England, Joint Information Systems Committee and Higher Education Academy. (2005) *HEFCE Strategy for e-Learning*. HEFCE; JISC; HEA. HEFCE 2005/12. [Online] Available from: http://webarchive.nationalarchives.gov.uk/20100202100434/http://www.hefce.ac.uk/pubs/hefce/2005/05_12/05_12.pdf.

Higher Education Funding Council for England; Online Learning Task Force. (2011) *Collaborate to Compete: seizing the opportunity of online learning for UK higher education.* HEFCE. HEFCE 2011/01. [Online] Available from: http://www.hefce.ac.uk/media/hefce1/pubs/hefce/2011/1101/11_01.pdf.

Higher Education Funding Council for Wales. (2010) *Corporate Strategy 2010/11-2012/13.* HEFCW. [Online] Available from: http://www.hefcw.ac.uk/documents/publications/corporate_ documents/Corporate%20strategy%202010%20-%20English.pdf.

Higher Education Funding Council for Wales. (2011) *Annual Report 2010-11.* HEFCW. [Online] Available from: http://www.hefcw.ac.uk/documents/publications/corporate_ documents/Annual%20Report/HEFCW%20Annual%20Report%202010-11%20English.pdf.

Higher Education Funding Council for Wales. (2013a) *Annual Report 2012-13.* HEFCW. [Online] Available from: http://www.hefcw.ac.uk/documents/publications/corporate_ documents/Annual%20Report/HEFCW%20annual%20report%202012-13%20English.pdf.

Higher Education Funding Council for Wales. (2013b) *Corporate Strategy 2013/14-2015/16.* HEFCW. [Online] Available from: http://www.hefcw.ac.uk/documents/publications/corporate_documents/Corporate%20Strategy%202013-14-2015-16.pdf.

Higher Education Funding Council for Wales. (2013c) *HEFCW Operational Plan 2012-13: Annex B.* HEFCW. HEFCW/12/33. [Online] Available from: http://www.hefcw.ac.uk/documents/publications/corporate_documents/operational%20plan%202012-13.pdf.

HM Government. (2009) *Archives for the 21st Century.* London: TSO. Cm 7744. [Online] Available from: http://www.nationalarchives.gov.uk/documents/information-management/archives-for-the-21st-century.pdf.

HM Government. (2013) *Information Economy Strategy.* London: BIS. [Online] Available from: https://www.gov.uk/government/uploads/system/uploads/attachment_data/file/206944/13-901-information-economy-strategy.pdf.

House of Commons; Culture, Media and Sport Committee. (2005) *Public Libraries: third report of Session 2004-05,* Volume 1. London: Stationery Office. HC 81. [Online] Available from: http://www.publications.parliament.uk/pa/cm200405/cmselect/cmcumeds/81/81i.pdf.

House of Lords; Science and Technology Committee. (2006) *Science and Heritage: Report with Evidence.* London: Stationery Office. HL Paper 256. [Online] Available from: http://www.publications.parliament.uk/pa/ld200506/ldselect/ldsctech/256/256.pdf.

House of Lords; Science and Technology Committee. (2007) *Science and Heritage: an update.* London: Stationery Office. HL Paper 168. [Online] Available from: http://www.publications.parliament.uk/pa/ld200607/ldselect/ldsctech/168/168.pdf.

Innovate UK. (2015) *Digital Economy Strategy 2015-2018.* Swindon: Technology Strategy Board. [Online] Available from: https://www.gov.uk/government/uploads/system/uploads/attachment_data/file/404743/Digital_Economy_Strategy_2015-18_Web_Final2.pdf.

Invest Northern Ireland. (2010) *Digital Northern Ireland 2020.* [Belfast]: n.k. [Online] Available from: http://secure.investni.com/static/library/invest-ni/documents/digital-northern-ireland-2020-report.pdf.

Jenkins, G. (2005) *Digitisation Policy and Strategy, National Library of Wales.* Aberystwyth: NLW. [Online] Available from: http://www.llgc.org.uk/fileadmin/documents/pdf/digitisationpolicyandstrategy2005_S.pdf.

Jisc. (1996) *Five Year Strategy 1996-2001.* [Online] Available from: http://www.webarchive.org.uk/wayback/archive/20100924221026/http://www.jisc.ac.uk/aboutus/strategy/strategy9601.aspx.

Jisc. (2001) *JISC Strategy 2001-2005.* [Online] Available from: http://www.webarchive.org.uk/wayback/archive/20100924221029/http://www.jisc.ac.uk/aboutus/strategy/strategy0105.aspx.

Jisc. (2005) *Partnership Boost to UK Higher and Further Education.* [Online] Available from: http://www.jisc.ac.uk/news/press-release-partnership-boost-to-uk-higher-and-further-education-13-jun-2005.

Jisc. (c2007) *Strategy 2007-2009.* [Online] Available from: http://www.webarchive.org.uk/wayback/archive/20100924221025/http://www.jisc.ac.uk/aboutus/strategy/strategy0709.aspx.

Jisc. (2008) *Digitisation Strategy.* [Online] Available from: https://www.jisc.ac.uk/whatwedo/programmes/digitisation/jisc_digitisation_strategy_2008.doc.

Jisc. (c2010) *Strategy 2010-2012.* [Online] Available from: http://www.webarchive.org.uk/wayback/archive/20100924221032/http://www.jisc.ac.uk/aboutus/strategy/strategy1012.aspx.

Jisc. (2012) *JISC Strategic Vision 2013-2017: the power and possibilities of digital enablement and digital futures.* [Online] Available from: https@//wiki.jisc.ac.uk/display/JISCtransition/Our+strategic+vision.

Jisc. (2013) *Jisc Strategy 2013-2016.* [Online] Available from: http://www.jisc.ac.uk/reports/jisc-strategy-2013-16.

Jisc. (c2014) *Our Catalogue of Services for Higher Education.* [Online] Available from: http://www.jisc.ac.uk/about/subscription-package-for-higher-education.

Jones, R. A. (2008) A marathon not a sprint: lessons learnt from the first decade of digitisation at the National Library of Wales. *Program.* Vol. 42, No. 2: 97–114.

Lang, B. (1998) Developing the Digital Library. In Carpenter, L., Shaw, S. and Prescott, A. (Eds.) *Towards the Digital Library: the British Library's Initiatives for Access Programme.* London: British Library, 227–233.

Leadership Foundation, National Union of Students, Higher Education Academy, Association for Learning Technology and Jisc. (c2014) *Changing the Learning Landscape: connect to the future: final report.* London: Leadership Foundation. [Online] Available from: http://www.lfhe.ac.uk/en/programmes-events/your-university/cll/index.cfm?utm_source=development&utm_campaign=cll.

Libraries NI. (2010) *Business Plan 2010-2011.* [Online] Available from: http://www.librariesni.org.uk/AboutUs/OurOrg/Business%20Plans/Business_Plan_2010-11.pdf.

Libraries NI. (2011) *Business Plan 2012-2015.* [Online] Available from: https://www.librariesni.org.uk/AboutUs/OurOrg/Corporate%20Plans/Corporate_Plan_2011-15.pdf.

Libraries NI. (2014) *Business Plan 2014-2015.* [Online] Available from: https://www.librariesni.org.uk/AboutUs/OurOrg/Business%20Plans/Business_Plan_2014-15_Final_Version.pdf.

Libraries NI. (2015) *Digitisation Policy.* [Online] Available from: https://www.librariesni.org.uk/AboutUs/OurOrg/Policies%20and%20Procedures/Digitisation_Policy.pdf.

Macdonald, L. (2012) *A New Chapter: public library services in the 21st century.* Dunfermline: Carnegie UK. [Online] Available from: http://www.carnegieuktrust.org.uk/CMSPages/GetFile.aspx?guid=b04629b2-aa09-4bd0-bc3a-9b9b04b7aba1.

Mahoney, J. (1998) Introduction. In Carpenter, L., Shaw, S. and Prescott, A. (Eds.) *Towards the Digital Library: the British Library's Initiatives for Access Programme.* London: British Library: 10–14.

Museums Association. (c2015) *Campaigns: government strategies.* [Online] Available from: http://www.museumsassociation.org/campaigns/government-strategy-and-support-for-museums#.U7FqgXxOUcA.

Museums Galleries Scotland. (2008) *Digital Content Action Framework.* Edinburgh: Museums Galleries Scotland. [Online] Available from: http://www.museumsgalleriesscotland.org.uk/research-and-resources/resources/publications/publication/3/digital-content-action-framework.

Museums Galleries Scotland. (2012) *Going Further: the national strategy for Scotland's museums and galleries*. Edinburgh: Museums Galleries Scotland. [Online] Available from: http://www.museumsgalleriesscotland.org.uk/research-and-resources/resources/publications/publication/460/going-further-the-national-strategy-for-scotlands-museums-and-galleries.

Museums Galleries Scotland. (2013) *From Strategy to Action: a delivery plan for Scotland's museums and galleries 2013-2015*. [Online] Available from: http://www.museumsgalleriesscotland.org.uk/research-and-resources/resources/publications/publication/504/from-strategy-to-action-a-delivery-plan-for-scotlands-museums-and-galleries.

Museums, Libraries and Archives Council. (2008a) *Corporate Plan 2008 to 2011*. London: MLA. [Online] Available from: http://webarchive.nationalarchives.gov.uk/20080603220646/http://mla.gov.uk/news/press_releases/cp.

Museums, Libraries and Archives Council. (2008b) *Framework for the Future: MLA action plan for public libraries − "Towards 2013"*. London: MLA. [Online] Available from: http://webarchive.nationalarchives.gov.uk/20080603220646/http://mla.gov.uk/news/press_releases/Five_year_plan_to_make_every_library_a_great_library.

Museums, Libraries and Archives Council. (2009) *Leading Museums: a vision and strategic action plan for English museums*. [Online] Available from: http://webarchive.nationalarchives.gov.uk/20091116103422/http://www.mla.gov.uk/what/strategies/~/media/Files/pdf/2009/MLA_Museum_ActionPlan_final.

Museums, Libraries and Archives Council. (2011) *Future Libraries Programme: final report to Governance Board*. [Online] Available from: http://www.local.gov.uk/c/document_library/get_file?uuid=70573292-bbde-4749-bedf-ff842a34d0a8&groupId=10180.

Museums, Libraries and Archives Council and Local Government Group. (2011) *Future Libraries: change options and how to get there, learning from the Future Libraries Programme Phase 1*. London: Local Government Group. [Online] Available from: http://www.local.gov.uk/c/document_library/get_file?uuid=72549e55-0b3e-44bb-af6c-8aa58590ad59&groupId=10180.

National Archives of Scotland. (1999-2007) *Annual Reports*. [Online] Available from: http://www.nas.gov.uk/about/annualReport.asp.

National Archives of Scotland. (2003-2013) *Corporate Plan[s]*. [Online] Available from: http://www.nas.gov.uk/about/plan.asp.

National Archives of Scotland. (n.d.) *Key Strategies*. [Online] Available from: http://www.nas.gov.uk/documents/keyStrategies.pdf.

National Library of Scotland. (2003) *Breaking through the Walls: the strategy of the National Library of Scotland*. [Online] Available from: http://www.nls.uk/media/22415/strategy2004.pdf.

National Library of Scotland. (2005) *Digital National Library of Scotland: Strategic Plan 2005-2008*. [Online] Available from: http://www.nls.uk/media/22406/nls_digital_library_strategy.pdf.

National Library of Scotland. (2008) *NLS strategy 2008-2011*. [Online] Available from: http://www.nls.uk/about-us/corporate-documents/2011-2014-strategy.

National Library of Scotland. (2010) *Thriving or Surviving? National Library of Scotland in 2030*. [Online] Available from: http://www.nls.uk/media/808985/future-national-libraries.pdf.

National Library of Scotland. (2011) *NLS strategy 2011-2014*. [Online] Available from: http://www.nls.uk/about-us/corporate-documents/nls-strategy-2008-2011.

National Library of Scotland. (2012) *Corporate Plan 2012-2015*. [Online] Available from: http://www.nls.uk/media/1024036/2012-2015-corporate-plan.pdf.

National Library of Scotland. (2013) *Corporate Plan 2013-2014*. [Online] Available from: http://www.nls.uk/media/1076420/2013-2014-corporate-plan.pdf.

National Library of Wales. (2007) *Shaping the Future: the Library's strategy 2008-2009 to 2010-2011.* Aberystwyth: National Library of Wales. [Online] Available from: http://www.llgc.org.uk/fileadmin/documents/pdf/nlw_strategy_s.pdf.

National Library of Wales. (2009) *Digitisation Strategy 2008/9 — 2010/11.* [Online] Available from: http://www.llgc.org.uk/fileadmin/fileadmin/docs_gwefan/amdanom_ni/dogfennaeth_gorfforaethol/dog_gorff_strat_dig_08_09_10_11S.pdf.

National Library of Wales. (2010) *Twenty-twenty: a long view of the National Library of Wales.* [Online] Available from: https://www.llgc.org.uk/fileadmin/fileadmin/docs_gwefan/amdanom_ni/dogfennaeth_gorfforaethol/dog_gorff_gwel_2020S.pdf.

National Library of Wales. (2011a) *The Agile Library: the Library's strategy 2011-2012 to 2013-2014.* [Aberystwyth]: National Library of Wales. [Online] Available from: http://www.llgc.org.uk/fileadmin/documents/pdf/Strategy2011-12_2013-14.pdf.

National Library of Wales. (2011b) *Digitisation Strategy 2011/12-14/15.* [Online] Available from: http://www.llgc.org.uk/fileadmin/fileadmin/docs_gwefan/amdanom_ni/dogfennaeth_gorfforaethol/StrategaethDdigido2012-2015.pdf.

National Library of Wales. (2014a) *Knowledge for all: National Library of Wales strategic plan 2014-2017.* Aberystwyth: National Library of Wales. [Online] Available from: http://www.llgc.org.uk/fileadmin/fileadmin/docs_gwefan/amdanom_ni/dogfennaeth_gorfforaethol/corff_strat_KnowledgeforAll_2014_2017S.pdf.

National Library of Wales. (2014b) *Research Programme in Digital Collections at NLW.* [Online] Available from: http://www.;;gc.org.uk/about-us/research/research/?no_cache=1.

National Library of Wales. (2015) *NLW Research>Projects.* [Online] Available from: https://www.llgc.org.uk/collections/activities/research/projects.

National Records of Scotland. (2008-2014) *Annual Reports and Accounts.* [Online] Available from: http://www.nrscotland.gov.uk/about-us.

National Records of Scotland. (2011) *NRS Business Plan 2011-2012.* [Online] Available from: http://www.nas.gov.uk/documents/NRSBusinessPlan2011-12.pdf.

National Records of Scotland. (2012) *Strategy 2012-2022.* [Online] Available from: http://www.nas.gov.uk/documents/NRSStrategy2012-2022.pdf.

Northern Ireland Digital Content Strategy 2012-2015: 'Moving Forward' (c2011). [Online] Available from: http://quayperformance.com/DC/Digital_Circle_Strategy_2012_-_2015.pdf.

Public Record Office of Northern Ireland. (2003) *Corporate Plan 2003 to 2006; Business Plan 2003/04.* [Online] Available from: http://www.proni.gov.uk/proni_corporate_plan_03-06_and_business_plan_03-04__pdf_557kb_.pdf.

Public Record Office of Northern Ireland. (2006-2014) *Deputy Keeper's Report.* [Online] Available from: http://www.proni.gov.uk/index/about_proni/who_are_we_and_what_we_do/business_plans_deputy_keepers_reports_and_annual_reports.htm.

Reading Agency (2013) *Library 21* [Online] Available from: http://theliteraryplatform.com/collective/2014/05/24/library-21-research-and-feasibility-study/.

Renaissance in the Regions: realising the vision: Renaissance in the Regions 2001-2008. (2009) [Online] Available from: http://webarchive.nationalarchives.gov.uk/20091116103422/http://www.mla.gov.uk/what/programmes/renaissance/ ~ /media/Files/pdf/2009/Renaissance_Review_Report.

Scottish Executive and One Scotland. (2006) *Scotland's Culture: Cultar na h-Alba.* Edinburgh: Scottish Executive. [Online] Available from: http://www.gov.scot/Resource/Doc/89659/0021549.pdf.

Scottish Executive and Scottish Library and Information Council (2007). *Building on Success: a public library quality improvement matrix for Scotland.* n.p.:SLIC. [Online] Available from: http://www.scottishlibraries.org/storage/activities/qse/plqim/plqim.pdf.

Scottish Funding Council. (c2011) *Universities: Putting the Learner at the Centre: Higher Education Outcome Agreements: Achievements and Ambitions.* Edinburgh: SFC. [Online] Available from: http://www.sfc.ac.uk/web/FILES/Funding_Outcome_Agreements_2014-15/Higher_Education_Outcome_Agreements_Achievements_and_Ambitions_2014-15.pdf.

Scottish Funding Council. (c2012a) *Colleges: Putting the Learner at the Centre: summary of progress and 2014-15 Outcome Agreements*. Edinburgh: SFC. [Online] Available from: http://www.sfc.ac.uk/web/FILES/Funding_Outcome_Agreements_2014-15/Colleges_Summary_of_progress_and_2014-15_outcome_agreements.pdf.

Scottish Funding Council. (2012b) *Scottish Funding Council Strategic Plan 2012-2015: delivering ambitious change*. Edinburgh: SFC. [Online] Available from: http://www.sfc.ac.uk/web/FILES/ReportsandPublications/strategic_plan_2012-15_web.pdf.

Scottish Government. (2011a) *Putting Learners at the Centre: delivering our ambitions for post-16 education*. Edinburgh: Scottish Government. [Online] at: http://www.scotland.gov.uk/Resource/Doc/357909/0120943.pdf.

Scottish Government. (2011b). *Scotland's Digital Future: a strategy for Scotland*. Edinburgh: Scottish Government. [Online] Available from: http://www.gov.scot/Publications/2011/03/04162416/0.

Scottish Government. (2012). *Scotland's Digital Future: delivery of public services*. Edinburgh: Scottish Government. [Online] Available from: http://www.gov.scot/Publications/2012/09/6272/0.

Scottish Government. (2013) *Scotland's Digital Future: supporting the transition to a world-leading digital economy*. Edinburgh: Scottish Government. [Online] Available from: http://www.gov.scot/Publications/2013/05/2347.

Scottish Library and Information Council. (2011a) *Operating Plan 2011-2014*. [Online] Available from: http://www.scottishlibraries.org/key-documents/.

Scottish Library and Information Council. (2011b) *Strategic Plan 2011-2014*. [Online] Available from: http://www.scottishlibraries.org/key-documents/.

Scottish Library and Information Council. (2014) *Business Plan 2014-2015*. [Online] Available from: http://scottishlibraries.org/wp-content/uploads/2015/02/SLIC_Business_Plan_-2014_2015_Final.pdf.

Scottish Library and Information Council. (2015) *Ambition and Opportunity: a strategy for public libraries in Scotland 2015-2020*. [Online] Available from: http://scottishlibraries.org/wp-content/uploads/2015/01/Strategy.pdf.

Scottish Museums Council. (2004) *A National ICT Strategy for Scotland's Museums*. Edinburgh: Scottish Museums Council. [Online] Available from: http://www.museumsgalleriesscotland.org.uk/research-and-resources/resources/publications/publication/40/a-national-ict-strategy-for-scotlands-museums.

Scottish Museums Council. (2005) *Museums, Galleries and Digitisation: current best practice and recommendations on measuring impact: summary report*. Edinburgh: Scottish Museums Council. [Online] Available from: http://www.museumsgalleriesscotland.org.uk/research-and-resources/resources/publications/publication/44/museums-galleries-and-digitisation-summary-report.

Smithies, R. (2011) *A Review of Research and Literature on Museums and Libraries*. London: ACE. [Online] Available from: http://www.artscouncil.org.uk/publication_archive/museums-and-libraries-research-review/.

The National Archives. (2007a) *Living Information: the Vision of The National Archives*. [Online] Available from: https://www.nationalarchives.gov.uk/documents/living-information.pdf.

The National Archives. (2007b) *Strategic Plan 2007 to 2008*. [Online] Available from: http://collections.europarchive.org/tna/20110131174836/http://www.nationalarchives.gov.uk/documents/strategic-plan0708.pdf.

The National Archives. (2008a) *Provide and Enable: The National Archives Online Strategy*. [Online] Available from: http://www.nationalarchives.gov.uk/documents/provide-enable.pdf.

The National Archives. (2008b) *Strategic Plan 2008 to 2009*. [Online] Available from: http://collections.europarchive.org/tna/20110131174836/http://www.nationalarchives.gov.uk/documents/strategic-plan0809.pdf.

The National Archives. (2009c) *Strategic Plan 2009 to 2010*. [Online] Available from: http://collections.europarchive.org/tna/20110131174836/http://www.nationalarchives.gov.uk/documents/strategic-plan0910.pdf.

The National Archives. (2010a) *See the Bigger Picture: our Plans for 2010-11*. [Online] Available from: http://collections.europarchive.org/tna/20110131174836/http://www.nationalarchives.gov.uk/documents/priorities2010-11.pdf.

The National Archives. (2010b) *Strategic Plan 2010 to 2011*. [Online] Available from: http://collections.europarchive.org/tna/20110131174836/http://www.nationalarchives.gov.uk/documents/strategic-plan10-11.pdf.

The National Archives. (2011) *For the Record. For Good: our Business Plan for 2011-2015*. [Online] Available from: http://www.nationalarchives.gov.uk/documents/the-national-archives-business-plan-2014-2015.pdf.

The National Archives. (2012a) *Archives for the 21st Century in Action: refreshed 2012-2015*. [Online] Available from: http://www.nationalarchives.gov.uk/documents/archives/archives21centuryrefreshed-final.pdf.

The National Archives. (2012b) *Digital Strategy*. [Online] Available from: http://www.nationalarchives.gov.uk/documents/the-national-archives-digital-strategy.pdf.

The National Archives and Museums, Libraries & Archives Council. (2010) *Archives for the 21st Century in Action*. [Online] Available from: http://www.nationalarchives.gov.uk/documents/information-management/archives-for-the-21st-century-in-action.pdf.

Wellcome Library. (2015a) *Digitisation at the Wellcome Library*. [Online] Available from: http://wellcomelibrary.org/what-we-do/digitisation/.

Wellcome Library. (2015b) *Transforming the Wellcome Library: 2009-2014*. [Online] Available from: http://wellcomelibrary.org/what-we-do/library-strategy-and-policy/transforming-the-wellcome-library/.

Wellcome Trust. (2003-2013) *Annual Report and Financial Statements [2003-2013]*. [Online] Available from: http://www.wellcome.ac.uk/About-us/Publications/Annual-Report-and-Financial-Statements/Previous/index.htm.

Wellcome Trust. (2005) *Strategic Plan: Making a Difference 2005-2010*. [Online] Available from: http://www.wellcome.ac.uk/stellent/groups/corporatesite/@policy_communications/documents/web_document/wtd018878.pdf.

Wellcome Trust. (2010) *Strategic Plan 2010-20: Extraordinary Opportunities*. [Online] Available from: http://www.wellcome.ac.uk/stellent/groups/corporatesite/@policy_communications/documents/web_document/WTDV027438.pdf.

Wellcome Trust. (2012) *Strategic Plan 2010-20: two-year update*. [Online] Available from: http://www.wellcome.ac.uk/stellent/groups/corporatesite/@policy_communications/documents/web_document/wtvm054497.pdf.

Wellcome Trust. (2014) *Annual Report and Financial Statements 2014*. [Online] Available from: http://www.wellcome.ac.uk/About-us/Publications/Annual-Report-and-Financial-Statements/.

Welsh Assembly Government. (2009) *Archives for the 21st Century*. [Online] Available from: http://gov.wales/docs/drah/publications/091203archives21stCenturyen.pdf.

Welsh Assembly Government. (2010) *Delivering a Digital Wales: the Welsh Assembly Government's Outline Framework for Action*. [Cardiff?]: Welsh Assembly Government. [Online] Available from: http://wales.gov.uk/docs/det/publications/101208digitalwalesen.pdf.

Welsh Assembly Government. (2011a) *Digital Wales: Delivery Plan: Delivering a Digital Wales*. [Cardiff?]: Welsh Assembly Government. [Online] Available from: http://wales.gov.uk/docs/det/publications/110427deliveryplan.pdf.

Welsh Assembly Government. (2011b) *Libraries inspire: the strategic development framework for Welsh libraries 2012-2016*. [Online] Available from: http://wales.gov.uk/topics/ cultureandsport/museumsarchiveslibraries/cymal/libraries/librariesinspire/?lang=en.

Welsh Assembly Government. (2013a) *Digital Wales: a review of delivery*. [Pontypridd?]: Welsh Assembly Government. [Online] Available from: http://wales.gov.uk/docs/det/ publications/130325dwdeliveryen.pdf.

Welsh Assembly Government. (2013b) *Digital Wales: Directory of Projects*. [Cardiff?]: Welsh Assembly Government. [Online] Available from: http://wales.gov.uk/docs/det/ publications/130426directoryofprojectsen.pdf.

Welsh Assembly Government and CyMAL. (2011) *Evaluation of Libraries for Life: summary report*. Edinburgh: Scotinform. [Online] Available from: http://wales.gov.uk/topics/ cultureandsport/museumsarchiveslibraries/cymal/researchandevidence/lflevaluation/ ?lang=en.

Welsh Government. (2011) *Concordat between The National Archives and Welsh Assembly Government*. [Online] Available from: http://gov.wales/about/organisationexplained/ intergovernmental/concordats/nationalarchives/?lang=en.

Welsh Government. (2014a) *Digital Wales: a review of delivery 2013-2014*. Pontypridd: Welsh Government, Digital Wales. [Online] Available from: http://gov.wales/docs/det/ publications/141128-review-of-delivery-en.pdf.

Welsh Government. (2014b) *Expert Review of Public Libraries in Wales 2014*. [Online] Available from: http://gov.wales/docs/drah/publications/141021libraries-review-report-en.pdf.

Welsh Government. (2014c) *Libraries making a difference: the fifth quality framework of Welsh Public Library Standards 2014-2017*. [Online] Available from: http://gov.wales/docs/ drah/publications/140425wpls5en.pdf.

CHAPTER 3

Digitisation Programmes and Outputs in the UK

Over the last 20 years, there has been much digitisation carried out in the UK, either by two or more organisations working in partnership or by individual institutions. Where public or philanthropic funding has been involved, the organisations' decisions about their digitisation work have been subject to the wider strategic context in which they each operate. As the previous chapter has shown, this has been and remains a very complex environment in the UK. Its many differences inevitably affect planning for digitisation, and even the one common factor in all the strategies examined, that is the commitment to collaboration, has several interpretations. Nonetheless, there have been many major initiatives in the UK since the 1990s, and while much remains to be done, the UK has delivered a very wide range of digitised content in very substantial quantities. Later chapters will discuss how future selection might be taken forward effectively to maximise further investments in digitisation. To inform those discussions, this chapter outlines the principal aims and the types of content delivered by exemplar programmes and providers, and how their strategies and choices have progressed as the organisations' experience of digitisation has matured. A listing of digitised outputs relating to each of the projects and organisations discussed in the following sections is provided in Appendix 1.

3.1 ELECTRONIC LIBRARIES PROGRAMME

The Electronic Libraries Programme (eLib) was one of the very first major programmes in the United Kingdom that contributed to the digitisation of analogue materials. It had its origins in the landmark *Follett Report* (Joint Funding Councils' Libraries Review Group, 1993). This review had been called to examine the increasing pressures and lack of space faced by academic libraries. Chapter seven of the *Report* discussed the concept of the 'virtual library', still a very new vision for the time, but one which inspired much interest and hope in the information world. The report made it clear that it was looking for ways to instigate significant shifts in library operations and thereby create an impetus for change across the whole higher education

Stepping Away from the Silos
ISBN 978-0-08-100278-0
http://dx.doi.org/10.1016/B978-0-08-100278-0.00003-9

sector. It set out to promote large-scale pilot and demonstrator projects focusing on the components of a putative electronic library service and to stimulate its creation. It also recognised that to develop the virtual library 'it will be necessary to have the printed material in electronic (digital) form, and it will also be necessary to have in place the electronic infrastructure for the delivery of the digitised material' (Joint Funding Councils' Libraries Review Group, 1993, para 274).

The consequent work was taken forward by the Follett Implementation Group on Information Technology (FIGIT), and the eLib Programme was born. Launched in 1994 with its first call for competitive bids, it ran in three stages: Phase 1 from 1995 to 1998; Phase 2 from 1996 to 1998, largely filling gaps identified in Phase 1; and Phase 3 from 1998 to 2001, consolidating and integrating related elements from the two earlier phases (Jisc, 1994, 1995, 1997). The response from the target communities was enthusiastic. There were 354 formal expressions of interest at the outset (Carr, 2007, p. 78). It was a truly innovative programme working in uncharted territory and aiming as the Programme Director subsequently wrote 'at a sea change, a cultural shift' (Rusbridge, 2001). Some 70 projects were supported over its lifetime, a deliberately large number of small projects to drive experimentation. While this approach was recognised by the formal evaluations to have some weaknesses (ESYS Consulting, 2000, 2001), it was widely acknowledged that eLib had 'clearly shaped current provision' (House, 2012, p. 48). It introduced the concept of the 'hybrid library', and a number of its projects grew into 'national information services on which UK research libraries now depend so much' (Carr, 2007, p. 34). These services included the Arts and Humanities Data Service (AHDS), the Higher Education Data Sharing Consortium (HEDS), Social Science Data Archive (SSDA), Resource Discovery Network (RDN), CURL Exemplars in Digital Archives (CEDARS) and the Digital Preservation Coalition (DPC). These in turn were amongst the building blocks for Jisc's Distributed National Electronic Resource (DNER). The full list of eLib projects is summarised in Appendix 1, and several fuller accounts of the programme, both contemporary and retrospective, have been published (Rusbridge, 2001; Green, 1997; Pinfield, 2004; Carr, 2007, pp. 73–90; Rikowski, 2011, pp. 68–71).

Given the extensive range of issues which FIGIT and eLib had to address, it was unsurprising that digitisation work *per se* emerged as only one of several strands. These covered access to network resources; digital preservation; electronic document delivery; electronic journals; electronic short-loan projects; images; large-scale resource discovery; on-demand publishing; preprints; quality assurance; supporting studies and training and awareness (Rusbridge, 1998; eLib, 2001). Ultimately there were only two projects which were specifically designated as addressing digitisation. Both focused on journal literature. The Digitisation in Art and Design (DIAD) project aimed to create a digitised record of selected in-copyright journals in the subject areas of design and applied arts to enable quicker access to the resources. The Internet Library of Early Journals concentrated on digitisation as a key mechanism to improve access to important research collections. As well as digitising 20-year runs of six classic, out-of-copyright titles from the 18th to 19th centuries, it also investigated some technical aspects which were fundamental to digitising older literature of this sort. Beyond the specific focus on digitisation in these two initiatives, a number of other eLib projects did also involve or relate to digitisation in some form. These included Digimap, Higher Education Library for Image eXchange (HELIX), CATRIONA II, Biz/ed, Edinburgh Engineering Virtual Library (EEVL), Institute of Historical Research-INFO (IHR-INFO), Resource for Urban Design Information (RUDI), and Medical Images: Digitised Reference Information Bank (MIDRIB).

3.2 NON-FORMULA FUNDING OF SPECIALISED RESEARCH COLLECTIONS IN THE HUMANITIES

Non-Formula Funding of Specialised Research Collections in the Humanities (NFF) was a parallel programme to eLib, funded jointly by all four Funding Councils, and ran in two phases from 1994 to 1999 (Parkinson and Bruce, c2000). Like eLib, it was developed to meet issues and opportunities identified in the *Follett Report* (Joint Funding Councils' Libraries Review Group, 1993). Its priorities were to enhance access to holdings of significant research value. It concentrated in particular on funding conservation, preservation and cataloguing of major research collections in academic institutions across the UK, as well as supporting publicity for the collections, and improved user services in relation to them. It also generated important innovative work on multilevel cross-searching of different archival formats (Mowat, 1998), which would later contribute to the establishment of the *Archives Hub*. The majority of the outcomes were in the form of catalogues, other finding aids, preservation, conservation and increased on-site access to the physical collections. At least 14 of the projects, however, included digitisation. Their outputs varied from the conversion of a set of documents as an innovative extension to cataloguing work, such as the *English Local History Collection*, or the *Gertrude Bell Archive*, through to digitisation as the core purpose of the project, for example, the *Aberdeen Bestiary* and the *Celtic Manuscripts (Bodleian and intercollegiate)*. Fuller details of all the NFF projects, including those covering digitisation, can be found in the summative guide to the NFF outputs (Parkinson and Bruce, c2000).

3.3 RESEARCH SUPPORT LIBRARIES PROGRAMME

The Research Support Libraries Programme (RSLP) was a further joint initiative of the four Higher Education Funding Councils that derived from the *Follett Report* (Joint Funding Councils' Libraries Review Group, 1993), and its subsequent investigation of research needs in the *Anderson Report* (Joint Funding Councils' Libraries Review Group, 1996). RSLP took a holistic approach to matters of research support, with the intention of combining traditional and new forms of access to research materials, to contribute to a 'Distributed National Collection' (*RSLP: Research Support Libraries Programme,* 2002; Milne, 2002). To do so, it operated in three strands: collaborative collection management; support for humanities and social science collections; and access, which included specially commissioned studies about means of facilitating access for researchers. The first two strands ultimately produced a range of bibliographic and archival records, collection-level descriptions, a modest amount of newly digitised content, and web directories and portals to lead users to these new resources. In total 53 projects were delivered, ten of which included digitisation of specific resources. Not unlike the NFF outputs, digitisation was a major part of some of the projects, such as *Charting the Nation*, and in other cases it was linked to wider works such as cataloguing, for example the *Charles Booth Archive*, or the creation of databases, directories and portals for access to the materials.

3.4 JISC
3.4.1 Distributed National Electronic Resource

Having taken responsibility for the eLib Programme after its inception under the auspices of FIGIT, Jisc became the Funding Councils' agency for successor projects in the higher education sector. The Distributed National Electronic Resource (DNER) ran from 2000 to 2003 and took forward many of the issues and concepts that had emerged from eLib (Rusbridge, 2001; Pinfield and Dempsey, 2001). At the same time, it was also responding to contemporary developments in the wider environment.

Some key technologies were beginning to stabilise and mature, while further innovations were being developed at an ever-increasing pace. As Pinfield and Dempsey (2001) put it in their account of the DNER's work, 'instead of information landscape what we have at the moment is more like 'information broadscape', and the DNER had been formulated 'to enable a coherent strategy to be developed for [the Jisc's] varying activities and services'. Its formally stated vision was to 'continue to provide an easily accessible, comprehensive information resource for use by learners, teachers and researchers within and beyond the UK higher education community' (Jisc, 1999). Crucially, it was expected to improve applicability of ICT to learning and teaching, eLib being perceived as relevant primarily to researchers and their interests. As the programme developed, it became clear that it was dealing in some cases with issues already identified at a local level by eLib, but now being addressed in a national context.

The initiative consisted of three complementary programmes to develop the DNER infrastructure; to create enhancements for learning and teaching in Jisc services content and digital libraries; and to evaluate the work done under these strands. By the end of the programme, some 54 projects had been funded. The initiative's interests were extensive, and digitisation work largely took place not as a separate exercise, but as a means to deliver projects focusing on adaptation of resources for learning and teaching materials; the provision and presentation of still images; portals; infrastructure services; interfaces for learners; resource discovery; assessment tools and techniques; and other elements similarly relevant to the development of the entire environment rather than specific content. In 2001 the DNER evolved into the Jisc Information Environment (Bruce & Notay, 2004), and as the projects completed their work, the various elements of the programme were repositioned in Jisc's strategy and developments. Amongst many new resulting initiatives was an independent focus on digitisation work.

3.4.2 Jisc Digitisation Programme 2004–09

The Jisc Digitisation Programme represented a major new development to meet the needs of UK higher education. It was responding not only to the lessons learned from earlier digitisation work, but also to government strategies for education which focused on exploiting IT and digital environments, and within which the need for digital content could be recognised (Department for Education and Skills, 2005; Becta, 2008). Digitisation was now 'part of Jisc's strategy to create new opportunities for learning, teaching and research through provision of authoritative and sustainable e-resources' (Glenaffric, 2009). The vision was to build sustainable and coherent e-resources, and this included making available 'hidden' collections (Jisc, 2006). The Digitisation Programme was run in two phases.

3.4.2.1 Phase 1: 2004—07

With this programme, the emphasis turned to large-scale digitisation, with six projects being funded. They were recognised as pioneering in their coverage of a variety of formats and subjects (Evidence Base, 2007, p. 13; Sykes, 2008). They included journals, film, population reports, newspapers and parliamentary publications. At the time they put Jisc amongst the leading organisations in the UK as well as internationally. The post-project evaluation reported that 'JISC (sic) has led the way in the UK and in terms of scale, subject coverage, cohesion and innovation compares favourably with the digitisation efforts of the international community' (Evidence Base, 2007, p. 15).

3.4.2.2 Phase 2: 2007—09

This phase of the Digitisation Programme built on the experience and lessons learned from the first round. It set out to provide online access for all users to a range of authoritative digitised resources. It emphasised in particular material that had previously been difficult or impossible to access. It aimed to provide digitised resources that were of broad disciplinary interest and that formed a coherent theme or themes. The call concentrated on the creation of a small number of large-scale and sustainable collections and/or the addition of material to existing collections to support learning, teaching and research. The 16 successful bids tackled further work on newspapers, journals and film, and other multidisciplinary resources including Cabinet Papers, 19th Century pamphlets, theses and ephemera. Several additional formats and media were introduced within these initiatives, such as cartoons, audio recordings and images.

3.4.3 Enriching Digital Resources 2008—09

This supplementary call addressed specific elements of the Jisc Digitisation Strategy (Jisc, 2008) as well as contemporary debates in the community, in particular about the need for reintegrating the user with the content delivered and to build not just a critical mass, but connected content. Twenty-five smaller projects were therefore funded, focusing less on creating new content than on better exploiting existing digitised resources.

The chosen projects were designed to meet one of three aims. The first was to carry out pilot and small-scale digitisation or smaller feasibility studies in preparation for larger scale work or to complete or add to existing resources where there were gaps or opportunities for expansion. The second was the enhancement of existing digitised collections that were underused or required extra development. The third covered the development of clusters of content by bringing together related digital resources. As the projects listed in Appendix 1 show, these aims also allowed digitisation of materials from a considerable variety of subjects and in various formats.

3.4.4 e-Content 2009–11

Following the close of the Digitisation Programme in 2009, the level of Jisc funding immediately available for further digitisation was limited, not least because of the pressures on higher education funding after the global economic downturn around 2008. Jisc recognised that, as well as there being demand for further large-scale digitisation, there were a number of related issues which should be investigated to build an effective environment for the creation and delivery of digital content. The e-Content Programme, therefore, did not focus on creating new digital content but on addressing two key elements of the planning and processes underlying digitisation. These were institutional skills and strategy; and clustering and enhancing digitised resources, to combat the 'silo' culture that was by then well recognised in the UK. Six of the projects explored the complicated and practical issues about achieving effective clustering of new content and content previously digitised with Jisc's support. The outcomes are summarised in a specially issued e-book (Jisc, 2011).

3.4.5 Content Programme 2011–13

This Programme was developed from the previous Jisc Digitisation and Content Programmes, in particular the two phases of the Programme from 2005 to 2009 described earlier, which had produced 20 major resources and helped to develop a critical mass of digital content from the UK's analogue collections. The Programme for 2011–13 supported 23 projects. By this period, there were several aspects of digitisation requiring further development, beyond the straightforward creation of content. The Programme therefore ran in three strands.

Strand A focused on digitisation and open educational resources (OERs), aiming to create OERs that incorporated scholarly material and were embedded within teaching and learning activities. A very broad range of content was invited, including image collections, manuscripts, sound recordings and moving images, maps, ephemera, primary data and newspapers. These could originate from libraries' special and archival collections and also from collections held in HE museums, or formed by departments or individual scholars. Nine projects were funded, with coverage ranging from medical science to historical and social subjects, design, architecture and dance.

Strand B covered eight mass digitisation projects, primarily meeting needs of higher education in the UK. Jisc aimed to digitise special collections and other analogue material of educational use, and to build sustainable and coherent digital resources that met the needs of higher education as well as being of interest at local, national and international level. Several of the funded projects covered various different fields of science, alongside historical and social materials.

Strand C, Clustering Digital Content, followed up Strand B of the e-Content Programme (2009—11) and responded to the importance of clustering and aggregating digital content to develop a more effective digital landscape for education. Seven projects were expected to build and expand services to bring existing digital content together for the benefit of, primarily, the higher education sector. The successful bids included both specific formats, such as manuscripts, maps, parliamentary records and broadside ballads, as well as resources for specific disciplines, including engineering, veterinary science and history.

3.4.6 Developing Community Content 2010-2011 and e-Content Programme 2011

Developing Community Content was not a digitisation programme in its purest sense but a wider concept that explored and developed ways that communities could develop their use of digital content within and for themselves as well as for wider audiences. The emphasis was on advancing the skills of communities to enable innovation in their use of digital information, co-create new resources and manage the various technical and legal aspects involved. An important deliverable of the project was the infokit that offers guidance on all these issues, with the individual projects serving as examples as well as actual information resources (Jisc Digital Media, n.d.). The outputs of these projects were an impressive range of resources that constituted newly created content, new content based on existing digital materials, some new digitisation or various combinations of these three elements.

3.4.7 Rapid Digitisation 2011

Financial limitations meant that by 2011 Jisc could not continue its earlier, large-scale digitisation programmes, but the organisation remained committed to supporting small-scale projects. The Rapid Digitisation Programme funded the creation of new content which was to be carried out in different ways and for different purposes. This included adding new content to existing resources, converting small collections, testing the viability and value of future large projects by pilot schemes, and digitising with new technologies. As the projects listed in Appendix 1 show, this was another programme that ranged widely in its coverage of subject and format, from 17th- and 18th-century manuscripts to music and metalwork.

3.4.8 UK Medical Heritage Library

With Jisc's translation from a committee of the joint higher education Funding Councils to an independent legal entity, all elements of the organisation's operations and activities were reviewed and remodelled, and during that time further major work on digitisation was on hold. Its first large-scale project since the change of status is the *UK Medical Heritage Library*, a partnership with the Wellcome Library and described more fully in this chapter's section on the Wellcome's activities. As indicated in Chapter 2, Jisc remains committed to supporting digitisation, but changes in its approach and priorities for further digitisation projects can be expected in future years.

3.5 RESEARCH COUNCILS

As outlined in Chapter 2, the UK's Research Councils (RCUK) have not placed emphasis on digital content creation or digitisation in their strategies, although digitised materials have been an integral element of the resources gathered for some projects across all disciplines. The exception is the Arts and Humanities Research Council (AHRC), and its predecessor the Arts and Humanities Research Board (AHRB). Their long-running strategic commitment to their Resource Enhancement Scheme reflected the importance of digital resources, whether born-digital or digitised from analogue materials, to research in Arts and Humanities, and specifically to the emerging discipline of Digital Humanities.

From 2000 the Resource Enhancement Scheme was one of two major AHRB/C funding streams which contributed actively to digital content creation, including digitisation. It was amongst the earliest AHRB programmes and was specifically intended to create and make accessible resources and scholarly information in digital form (Dunning, 2010, p. 119). It ran from 2000 to 2007, and supported 173 projects (see Appendix 1) all focused on producing digital research resources (Robey, 2008). It functioned in parallel to the AHRB/C Standard Research Grants (SRG) programme, which supported mainstream research and its outputs in all forms. A substantial number of projects from this funding stream also included the creation of digital research resources, albeit as ancillary elements (Robey, 2008). As a result, in the period 1999—2009 the AHRB/C funded by far more digital research resources than any other UK public funder or charity (Winterbottom and Robey, c2010).

The Resource Enhancement Scheme ceased in 2007, following a formal review in 2005. The issues that led to the programme's demise have been outlined by the AHRC's Director of Research (Llewellyn, 2012), and Dunning notes in particular that the programme's review found it to be ineffective in identifying and meeting gaps in resources required for the Arts and Humanities community (Dunning, 2010, p. 119). This did not, however, signal the Council's withdrawal from supporting digitisation, as they continued to fund the creation of digital research resources under the ongoing SRG programme (Robey, 2008). In the following years, the AHRC updated its strategy and funding programmes to align with the RCUK's overall policy of addressing crucial issues with societal and economic impact and, within these, the RCUK Digital Economy Research Programme. It was not surprising that digital content creation and digitisation dropped in the overall priorities and, when supported, was subsumed into wider projects. Nonetheless, more recent programmes could and did include further digital content creation and digitisation as an integral part of the research work and its outputs. Currently, the Council's funding initiatives are driven by their four overarching themes, which include the Digital Transformations Theme. Its central goal is 'the *transformation* brought about by incorporating and using the digital in arts and humanities research rather than on the digital itself' (Llewellyn, 2012). While this might seem the most likely home for further digitisation, funding for resource enhancement *per se* is typically not awarded, but may be supported where a strong research question requires it to explore the research issues (AHRC, c2011). In the current climate, this would seem to be the most likely basis for any further digitisation work to be pursued under the Council's auspices.

3.6 NEW OPPORTUNITIES FUND

The New Opportunities Fund (NOF) was a major initiative of the UK Government, launched in the late 1990s, to support their strategy outlined in Chapter 2. It addressed fundamental issues, and proposed the establishment of the People's Network, which was to bring network access to public libraries and therefore within the reach of the entire population.

NOF also had a number of other strands, including its ICT Content Programmes, which were the largest such publicly funded programmes in Europe (Education for Change, 2006, p. 12). A major element of these was the programme Digitisation for Learning Materials, subsequently known as nof-digitise or NOF-Digi. Running from 1999 to 2004, the programme aimed to digitise existing material that could be used as educational content, and also to integrate new materials that would enhance the learning value of the electronic resources and content management, website design and marketing. The emphasis was very much on issues found in the government strategies of the time: social inclusion, equality, diversity, supporting the disadvantaged and promoting lifelong learning (Woodhouse, 2001; Education for Change, 2006; Big Lottery Fund, 2006).

This programme was complementary to similar initiatives for specific sectors, such as eLib and its successors, described elsewhere in this chapter. It was, however, available to a much wider constituency which included public libraries, museums, archives, galleries and other institutions in the public, private and voluntary sector. It had three overarching themes: cultural enrichment, citizenship in a modern state and reskilling the nation. It aimed to fund content that reflected the rich diversity of resources available, in particular through partnership between different organisations (Woodhouse, 2001). From 343 expressions of interest, 147 projects were funded over three years. Of these more than 80% were for digitisation of 'cultural enrichment content', with about 18% awarded for 'citizenship in a modern state' and 10% for 'reskilling the nation'. Digitisation, therefore, proved to be the area of most interest to participants. These came from the entire spectrum of educational and heritage organisations, with input from 74 libraries, 54 museums, 53 archives, 20 educational institutions and 106 others covering art, archaeology, architecture, environmental, health and general community groups. The scale of the projects varied widely (Education for Change, 2006) and their coverage was equally wide-ranging, 'on topics as diverse as biscuits, voluntary work, migration, biodiversity, football, contemporary art, music and photography, reading, etc…' (Woodhouse, 2001).

NOF-Digi was a groundbreaking development for all the sectors involved, and notably for public libraries. It was 'a major catalyst in giving many public libraries a first taste of digitization and digital content creation' (McMenemy, 2009, p. 120). It had some noteworthy successes, but its impact was also affected by some of the key developmental issues about digitisation that were part of the learning experience of all early projects. 'Digitization was a complex problem that needed much sophisticated strategic thinking before there could be a serious injection of further public funds' (Dunning, 2010, p. 119). Perhaps the most prominent and enduring issue has been the sustainability and continuing availability of the content created (Poole, 2009; Dunning, 2011).

3.7 BRITISH LIBRARY

The British Library was amongst the first major institutions in the UK to adopt the new digital technologies of the 1990s with a range of ground-breaking initiatives. Up to the year 2001/2 'the individual projects were worthy in their own right but there was no strategic approach or direction'. This, however, 'reflected the Library's desire to gain a body of knowledge on which to build a future strategy' (National Audit Office, 2004, p. 14). This situation was typical of early digitisation work in large institutions and the Library's first digital projects played a significant role in the developing field. They numbered around 20 and were brought together in 1994 to form the Initiatives for Access (IfA) project. Although much smaller than eLib, this was analogous to the higher education programme, in that it aimed to learn about new ways of working for the benefit of services and users, to produce demonstrators, and to engender enthusiasm for and understanding of the new technologies in the Library's staff (Mahoney, 1998, p. 11).

IfA ran until 1998 and was considered a highly successful venture. It explored the many central aspects of the new digital developments, including technical and infrastructural matters, text storage, cataloguing and other discovery methods, document delivery and textual analysis. It also included some early digitisation projects that proved to be very important both in establishing standards and good practice, and in creating the first building blocks for the Library's now long-established strategic ambition to create a critical mass of digitised content. As described in Chapter 2, from 2000 onwards, the Library accelerated its digitisation, digital collecting and digital archiving. Its *Annual Reports* (British Library, n.d.), and the Action

Plans within them, recount the development of its full digitisation strategy, issued in 2008 (British Library, 2008). As Appendix 1 shows, the Library's outputs are exceptionally wide-ranging in their coverage and vary from mainstream, large-scale projects to the small and highly specialised. This reflects the breadth and wealth of the Library's collections, the Library's strategic priorities for digitisation, and in some cases the specific interests of the Library's many partners and funders. Collaboration has been at the core of much of the Library's digitisation work, to maximise all aspects of the investment and outcomes, and to respect the steer of the House of Commons Culture Media and Sport Committee in 2000, which encouraged digitisation but not at the expense of the Library's core statutory functions (House of Commons; Select Committee on Culture, Media and Sport, 2000, para 86).

Within the extensive range of digitised content produced by the Library, there are many major initiatives that reflect careful strategic planning, and in some instances, as the *Annual Reports* show, its execution in stages over many years (British Library, 1999—2015). Amongst earliest outputs from IfA were some resources that have continued to be developed and have become central to the Library's digitisation strategy and provision. The first Treasures converted in the late 1990s (and presented using the awarding-winning technology of Turning the Pages) included *Beowulf*, the *Lindisfarne Gospels*, the *Sforza Hours*, and the *Notebook of Leonardo da Vinci*. Over time these have been joined by many more world-renowned items, such as the *Luttrell Psalter*, the *Gutenberg Bible*, Caxton's printing of Chaucer's *Canterbury Tales* and the *Sherborne Missal*. *Images Online*, now a major source of visual materials, started around 2000 with a project to put 100,000 items online by 2003. In the same period, the Library was digitising its contributions to the *International Dunhuang Project*. This collaborative venture involving institutions from Europe and Asia has reunified material in many formats from the Dunhuang Caves and the Eastern Silk Road, and is now a globally renowned resource (Shenton, 2009, pp. 39—40). Turning to a modern medium, in its most recent *Annual Report* at the time of writing (British Library, 1999—2015, see 2014—15), the Library issues a serious warning about the future of sound collections; it is generally acknowledged in the archival sector that they could be lost within 15 years, due to physical degradation and lack of equipment to play them. In response to this, the Library has initiated *Save our Sounds,* a major new programme launched in 2015. It will contribute to further development of *British Library Sounds*, the digitised collections that started around 2001 as the National Sound Archive and that will be an important part of the Library's plans to preserve its sound collections and ensure access for posterity.

Amongst the most prominent of the Library's digitisation initiatives has been the strategic goal to secure the long-term access to and preservation of its newspaper collections, which span three centuries and include 52,000 local, regional, national and international titles. Once again, the IfA Programme gave first attention to this material, with its work on the *Burney Collection of Early English Newspapers*, covering the 17th and 18th centuries. By 2003, a plan was in place to give researchers fully searchable access to digitised newspapers, and several major, large-scale programmes over the years since then have progressed systematically to create critical mass in the form of the *British Newspaper Archive*, developed in partnership with findmypast and launched in 2011. The work continues, with a new, ten-year plan initiated in the same year to extend further the online coverage of this important resource (British Library, 1999—2015, see 2011—12, p. 9).

By the middle of the last decade, the Library's activity in digitisation continued to reflect the breadth of its collections and of its potential to advance digitisation in many very different areas. In 2004, work began on the *Codex Sinaiticus*, the earliest surviving Christian Bible (British Library, 1999—2015, see 2008—09, p. 10; Shenton, 2009, pp. 33—34), with an overwhelmingly successful launch of its website in June 2008. The entire project was completed in 2009. Also in 2004, the Library became the administrator for the *Endangered Archives Programme*, dispensing grants from the Arcadia Fund to preserve, to date, endangered records in some 78 countries and safely relocate them both within the local region and on the British Library website (Shenton, 2009, pp. 42—43). In 2006, an agreement with Microsoft began the digitisation of 100,000 out-of-copyright books from the 19th century, covering both the works of famous authors and forgotten literary gems. Further mass digitisation of 250,000 books, pamphlets and early periodicals, was started in 2011 under an important partnership with Google to run over some five years. Digitisation of UK theses was also taken forward, with the launch in 2008 of the *Electronic Theses Online Service* (*EThOS*).

The most recent major initiatives by the Library continue to reflect its long-term commitment to digitisation, and to meet the Library's overall strategy in doing so. Leading examples include the development of the *Qatar Digital Library*, which started in 2010, in partnership with the Qatar Foundation. India Office records relating to the Arabian Gulf and Arabic manuscripts are being made available online, and the project has now entered a second phase from 2015 to be completed in 2018. Starting in 2011, the Library has contributed 10,000 items to *Europeana 1914–18*, with 500 of them also available on its own dedicated website, entitled *World War One*. Other prominent initiatives in the last two years include the *Hebrew Manuscript Digitisation*, a three-year project that has already been extended to 2019 by a new agreement with the National Library of Israel, and the *British in India*, the digitisation of 2.5 million microfilm records of the lives of the British in India from 1698 to 1947.

The briefest glance at Appendix 1 shows that the above programmes are only a fraction of the Library's overall output, but they reflect both the strategic potential and the complexities of planning digitisation in a large, world-class institution.

3.8 NATIONAL LIBRARY OF SCOTLAND

The National Library of Scotland's commitment to digitisation, as outlined in Chapter 2, was based on the Library's work as another early adopter in the 1990s. Its *Annual Reviews and Reports* from 1997 onwards describe major directions and initiatives from this period to the present day (NLS, 1997–2014). Even from the very early activity, its outputs included both materials where the digitised product was the end in itself, and also items for specific use as educational resources. These early OERs were originally developed in collaboration with the Scottish Consultative Council on the Curriculum, and disseminated either through the Scottish Virtual Teachers Centre website and on the Library's Resources for Learning in Scotland (RLS) site (now the online Learning Zone). *Annual Reviews* from 1997 to 2005 show that prominent from the start was the digitisation of the manuscript maps by Timothy Pont, the first stage in a series of projects to expose the Library's leading collections of maps from the 16th century to the present day. Alongside these came digital versions of World War I materials; local studies resources; medieval treasures; rare and unique books from before 1701; works of Burns and other leading literary figures; material reflecting the influence of Scots abroad; classic resources such as the first and second editions of the *Statistical Account of Scotland*; Shakespeare Quartos; music and songs; and early photographs.

This range of work was an important element in the foundations of the institution's strategy for the Digital National Library of Scotland, 2005-2008 (NLS, 2005).While this document covered the Library's plans for all aspects of digital provision, digitisation was presented as the first of the strategy's objectives and it was planned, subject to funding, to scale up activity in 2007 'to allow digitisation of entire collections for researchers' (NLS, 2005, p. 10). This strong commitment was reinforced in the Annual Review for 2005/6 which stated that 'the development of the digital library is now one of the key gateways to our great treasures and resources' (NLS, 1997–2014, 2005/6, p. 2). The Library's Integrated Collecting Strategy of 2008 committed to reallocation of financial resources to support digital developments (NLS, 2008, p. 4).

Digitisation of NLS materials since 2005 reflects this ambition to increase both the Library's activity and the quantities of materials digitised. *Annual Reviews* up to the time of writing highlight ongoing growth of types of material and subject areas already addressed and major new initiatives (NLS, 1997–2014). The latter include digitisation of thousands of Gaelic books; Scottish Post Office Directories; materials on the medical history of British India; and films from the Scottish Screen Archive, which became part of the National Library of Scotland in 2007. As of early 2015, the Library's online Digital Gallery presented its digitised collections or major resources with the wide range of subjects given broad subject classifications that covered art and design; films; kings and queens; literature and writers; maps and mapping; medieval manuscripts; music and song; photography; poetry; printing, publishing and the history of the book; religion; Scottish history and people; Scottish places; social history; sport; theatre; travel and emigration; and war.

3.9 NATIONAL LIBRARY OF WALES

As described in Chapter 2, digitisation has been a central plank of the National Library of Wales' strategy since 2001. This strong commitment at policy level has been borne out in the Library's outputs since that time.

Like other major institutions, the Library first experimented with digitisation in the mid to late 1990s. As Jones describes in some detail, its early work concentrated on online exhibitions and digital versions of selected treasures from the Library's collections (Jones, 2008). This led in 1999 to the initiation of four pilot projects; a learning resource based on an earlier exhibition about David Lloyd George; *Ymgyrchi! Campaign! ¡Campaña!*, an interpretive site about themes of Welsh politics; *Framed Works of Art*; and the *Dictionary of Welsh Biography*. The first of four digitisation strategies to date followed on these initiatives in 2000 (Green, 2002; Jones, 2008), along with more pilot projects, including the

Illingworth Cartoon Project. Most significantly, the Library's first official digitisation programme was developed, to include the *Treasures Programme*, the *Probate Project*, *Pre-1900 Welsh Journals,* the *John Thomas Photographic Collection* and the first phase of the *Geoff Charles Collection*. Momentum increased steadily in this period, leading to the Digitisation Policy and Strategy of 2005 (Jenkins, 2005), which drove further development of categories that were already partially digitised, as well as ventures into new areas. Thus work progressed on *Image Wales, Portrait Wales, Lloyd George Online, Words from the Past* (covering Welsh printed books and manuscripts), *the Family and Community History Resource Centre, the Welsh Abroad* and *Exhibiting Wales*.

By 2008, the Library was well positioned to move 'from iconic cherry-picking digitisation to large-scale if not mass digitisation' (Jones, 2008). Its new *Digitisation Strategy* (NLW, 2009) aimed to create, over time, a critical mass of content. This was reinforced by the instigation of a formal strategy to digitise the total printed record of Wales and the Welsh. To be called the *Theatre of Memory: Welsh Print Online* (NLW, 2011), it was to be as comprehensive as possible, regardless of the content's subject, genre, format or intellectual level and was to be freely available on the Internet. Within this, the 'most ambitious digitisation project ever undertaken by the Library' (National Library of Wales, 2012, p. 5), *Welsh Journals Online* and *Welsh Newspapers Online* were major contributions to the new concept, alongside other large collections such as the *Dictionary of Welsh Biography, Welsh Ballads Online, the Wales-Ohio site, Glaniad* (on the Welsh communities in Patagonia), *Portread (Portrait), Parish Registers* and the wide range of smaller collections and individual items digitised in the earlier years of the Library's digitisation activities. Since the start of this initiative, the Library has continued to pursue major new digitisation work, with more recent projects including *Cymru 1914: the Welsh experience of the First World War*, and *Cynefin*, a collaboration led by Archives Wales to digitise Welsh tithe maps, using crowdsourcing for transcription and community involvement. The Library has also contributed to the Imperial War Museum's site *Their Past Your Future*, covering the end of World War II and subsequent major conflicts.

As well as providing access to Welsh materials within the National Library, the institution's digitised materials have also made major contributions to the *People's Collection Wales*, part of the Welsh Assembly Government's 'One Wales' programme and intended to provide access to materials on Wales in all formats through a single website. The *People's*

Collection Wales is run in partnership by the National Library of Wales, the National Museum of Wales and the Royal Commission on Archives and Historical Manuscripts Wales (Tedd, 2011).

3.10 THE NATIONAL ARCHIVES

The strategic position of The National Archives (TNA) is particularly complex, given the organisation's responsibilities for supporting the business of government and advancing education at all levels, as well as curating and enriching the wider heritage of England and Wales and leading the archives sector in those two countries. Its strategy and planning over the years, as outlined in Chapter 2, reflect the many facets of this position, but make very clear that the digital environment in its widest sense is a major factor in the organisation's long-term vision and services. Within this, it is recognised that 'digitisation is a huge area of opportunity for archives' (TNA, 2012).

By 2002, TNA was firmly committed to progressing major digitisation projects, and by 2007 it planned to 'grow our digitised resources, both pre-digitised and through digitisation on-demand, so that over 90% of what customers request can be located and delivered remotely' (TNA, 2007, p. 13). Two formal digitisation programmes recorded the resources to be digitised and the progress on their delivery (TNA, 2005, 2009). These show clearly a range of systematic activity that, over the last 10 years or so, has delivered online access to core resources such as the Censuses, and to an extensive set of related materials from military, naval, air force and other historical sources. At the time of writing, TNA's website gave access alongside its general and military categories to selected materials that record the country's population and its development over the last two centuries. These range across many themes, including politics, art and design, crime, history, maps, taxation and wills and probate (TNA, 2015a, 2015d).

In 2004 the organisation decided to use the Licensed Internet Associates programme to fund digitisation. Non-exclusive licensing agreements with trusted commercial partners have enabled digitisation of major resources including the Censuses and several others of particular interest to family historians and genealogists (Griffiths & Maron, 2009; Maron, 2011). Access is free on TNA's premises, but is subject to a charge for online use from external locations. At the time of writing, such licensing is a well-established method and continues to be the mainstay of TNA's strategy for digitisation (TNA, 2015c). As a result, by 2009, some 90% of TNA's digitisation work had been created in partnership with commercial providers, and of that, some 90% in turn relates to family history, with the remaining 10% covering academic, military and other records (Griffiths & Maron, 2009).

Overall, at the time of writing, TNA has digitised and published online over 80 million of its historical documents (TNA, 2015b), representing approximately 9% of its total holdings (Griffiths & Maron, 2009). Based on its strategy as understood in early 2015, it can be presumed that this total will continue to rise over a period of many years, but probably with a minor proportion of new content being produced through public or philanthropic funding.

3.11 NATIONAL RECORDS OF SCOTLAND

The National Records of Scotland (NRS) and its predecessors, the National Archives of Scotland (NAS) and the General Register Office for Scotland (GROS) have been very active in digitisation. Both of the earlier organisations entered the field in the late 1990s, often in close partnership with each other to create resources that are now key elements of NRS provision.

One of the earliest initiatives that digitised archive materials both from the NAS and from some 52 local archive services was the *Scottish Archive Network* or SCAN. This project ran from 2000 to 2003. As well as providing catalogues, reference services, virtual exhibitions and resources for schools, it digitised a wide range of selected local materials, many of them quite small but, significantly, it also instigated the conversion of all Scottish wills from 1500 to 1925, an early example of large-scale digitisation, in collaboration with the Genealogical Society of Utah. By 2005 there were some 3 million pages completed, and these were combined with the GROS records of births, marriages and deaths in Scotland to form the backbone of *ScotlandsPeople*. This new venture was launched in 2005 to provide an enhanced and seamless service for all interested in Scottish family history (NAS, 2005, p. 3). A further agreement in 2002–03 with their partners from Utah (NAS, 2003, p. 30) led to digitisation of the Kirk Session records held in the NAS. By 2005-06, 2 million pages (or 80%) of these records were available online (NAS, 2006, p. 4) and were in due course added to *ScotlandsPeople*, making the activities of the Church of Scotland another of the principal resources on the site.

Other initiatives involving the NAS from about 2000 onwards included early contributions to *SCRAN*[1] (the subscription service providing resources

[1] http://www.scran.ac.uk.

for studying heritage and material culture) (NAS, 2000, p. 21), and, with NOF funding, the development of RLS in partnership with the National Library of Scotland and SCRAN. This included digitisation of approximately 1700 selected records from NAS collections (NAS, 2004, p. 50). Around the same time, the NAS was a key partner in the digitisation of the *Statistical Accounts of Scotland*, with funding from six other partners, mostly based in Scotland (NAS, 2001, p. 22). The NAS also collaborated in several other projects, including *Charting the Nation*, which was led by the University of Edinburgh with RSLP funding and which focused on digitising maps of Scotland and associated texts.

In addition to the NAS' early work on SCAN, this period saw the digitisation of wills, the creation of online learning resources, and from 2002-03, onwards a complementary 18-month project to create Scottish Archives for Schools (SAfS).[2] Developed in partnership with the Scottish Executive's Education Department, and Learning and Teaching Scotland, this initiative concentrated on Scottish history, and continued after the defined project period, with the service being extended in 2006-07 (NAS, 2007, p. 21) and continuing to develop to the present day.

In 2004-05 another significant initiative was undertaken by NAS and GROS. The Register Archives Conversion Project (RAC) took forward the mass digitisation of the *Register of Sasines* (NAS, 2005, p. 16, 2006, p. 28). As legal documents recording the transfer of land and building ownership, sasines represent another important component in the social and family history of Scotland. This major project ran over several years (NAS, 2011, p. 65).

In 2009 *ScotlandsPlaces* was launched, and formed a complementary service to that of *ScotlandsPeople* in particular (NAS, 2010, pp. 8—9). This collaboration between the NAS and the Royal Commission on the Ancient and Historical Monuments of Scotland focused on the history and heritage of places in Scotland. It brought together resources from a variety of government-funded organisations, and following its initial success the site was further developed, with the National Library of Scotland also joining the partnership.

3.12 PUBLIC RECORD OFFICE OF NORTHERN IRELAND

With the active commitment in 2003 to a planned digitisation programme, the Public Record Office of Northern Ireland (PRONI) steadily converted

---

[2] http://www.scottisharchivesforschools.org.

genealogical resources as anticipated, in keeping with the growing interest in this area that other UK archives had also identified. PRONI did not, however, limit its activities to this area. Digitisation work up to the present day has addressed a wider range of interests, including photographs and images from glass plate negatives of relevance to the local communities, Belfast maps, Londonderry City Council Minute Books, and personal documents relating to topics as varied as aviation in Northern Ireland and travel in Asia. Most recent initiatives have also included digitisation of local materials from the First World War (PRONI, 2006–2014).

3.13 WELLCOME LIBRARY

Chapter 2 outlines the Wellcome Trust's long-term commitment to addressing major global issues relating to the improvement of human and animal health. Its support extends actively to sustaining the types of infrastructure and resources necessary to support world-class research and advances in their fields of interest. It was in this context that the Wellcome Library made a formal commitment in 2007 to a major digitisation pro-gramme for its collections (Wellcome Library, 2015a), which was under-pinned by its *Transformational Strategy 2009–2014* (Wellcome Library, 2015c).

While this was a step change in strategic terms for the Wellcome Library, it was by no means a new direction for the organisation. The Library had already been involved in significant digitisation activities for some ten years at least, with its then Librarian stating as far back as 2001 that it was 'committed to creating a critical mass of digital surrogates based on its print holdings' (Pearson, 2001). Since that date its activities had included key mass digitisation projects, a prime example being *Medical Journal Backfiles*. This project was a collaboration with Jisc and the US National Library of Medicine (NLM) and it aimed to digitise a selection of medical journals, some dating back to the early 19th century. By 2005, over one million articles comprising some two million pages were to be available for inclusion in the *PubMed* database, with contributing publishers providing current issues (Jisc, 2013). The second stage of the project ran from 2006 to 2010 and added complete back issues covering nearly 200 years of histor-ically significant biomedical journals (Wellcome Trust, 2006). Another example of the Library's work in this period was the *Wellcome Film* project. This three-year programme digitised over 100 hours of film and video in the Library's collections. By 2009 its output was becoming available online, and this now forms part of the Wellcome's *Film and Sound Archive* (Henshaw, 2009; Wellcome Trust, 2009).

In the first decade of the 21st century, the Wellcome Trust gave active support to organisations such as archives, libraries, record offices, and scientific and clinical institutions where materials relevant to the Wellcome's interests were held. Research Resources in Medical History (RRMH) (Wellcome Trust, 2012), its formal scheme to make such resources more widely available, was first conceived in 1999 and implemented in 2001 in collaboration with the British Library, originally for a limited period. Its success led to its continuation after an initial review. At that time applications were mostly for cataloguing, preservation and conservation of printed and archival material, but the programme also recognised the value of digitisation, and a substantial number of the projects funded up to 2011 included digitisation, either as the principal activity or as one part of a wider programme. These are itemised in Appendix 1.

A further review of the RRMH programme in 2009 led to future funding applications being linked to the Wellcome's five major challenges, and the Wellcome Library's strategic shift to digitising 'a substantial portion of its holdings and making content freely available on the web' to become a world-class online resource for the history of medicine (Wellcome Library, 2015a). In *Transforming the Wellcome Library*, the vision is 'to be the pre-eminent destination — physical and online — for anyone interested in exploring health in its cultural and historical contexts' (Wellcome Library, 2015c), and one of its three core activities is strategic digitisation. The aim is to digitise as much of the Library's physical holdings as possible, and also to support digitisation and hosting of collections from partners that complement Wellcome holdings. The express intention is to 'permanently break the bonds imposed by the physical library and provide full access to…collections in new and innovative ways' (Henshaw & Kiley, 2013). In 2010, the target was 30 million digitised pages to be completed over five years (Henshaw et al., 2010), and this has been extended and increased more recently to 50 million pages, to be freely available by 2020 (Wellcome Trust, 2014a).

The year 2010 also saw the Trust's Board of Governors approve a multimillion pound digitisation programme, setting the stage for a long-term initiative (Henshaw et al., 2010, p. 53). The first phase of this new approach focused on Modern Genetics and its Foundations (Wellcome Trust, 2010). A two-year pilot project was based on Wellcome collections and the initiative was then widened by partnerships in the UK to make available complementary archive holdings of other institutions. Its coverage ranged from 1400 books on genetics and heredity published from 1850 to 1990 to the archives and papers of leading geneticists including Nobel-prize-winning scientists Francis Crick, Fred Sanger and Peter Medawar. Launched in 2013 as *Codebreakers: Makers of Modern Genetics* (Henshaw & Kiley, 2013; Wellcome Trust, 2013b), this venture also generated additional projects, which included digitisation of the Medical Officer of Health Reports for Greater London, available since 2013 on the *London's Pulse* website (Wellcome Trust, 2013a).

With the digitisation programme linked to Wellcome's research challenge areas, forthcoming themes can be expected to reflect this approach for some time to come. However, the funding criteria have remained very flexible, to accommodate applications that might not immediately fall within the core criteria of the programme, yet are relevant to the themes under development. Further work has also been progressed to add to and complement previous digitisation outputs (Wellcome Library, 2015b). In 2013, funding of £5.8 million was made available to focus on neurology and mental health (Wellcome Trust, 2003–2013, see 2013, p. 17). The following year Wellcome formed a major new partnership for a two-year project with archives in London and York, as well as leading psychiatric hospitals, to digitise 800,000 pages of medical and related archival materials covering from the 18th to 20th centuries (Wellcome Trust, 2014a).

In 2014, another major programme capitalised on the *Medical Journals Backfiles Project* that ran from 2004 to 2010. Wellcome and the National Library of Medicine (NLM) in the United States agreed a three-year project to convert complete back issues of historically significant biomedical journals. NLM journals covering 150 years are to be scanned at article level, with the output to be available through *PubMed Central* and *Europe PMC*. Part of the project will concentrate on mental health journals, in keeping with the Wellcome's major archive digitisation programme described earlier (Wellcome Trust, 2014b).

Also initiated in 2014 was the *19th Century Medical Books Digitisation Project* (Wellcome Trust, 2014c, 2014d), another programme which will become part of the *UK Medical Heritage Library*, first developed by Wellcome in 2013 and linked to the international Medical Health Library. In a significant partnership with Jisc and nine other institutions, including universities and Royal Colleges of Medicine, c15 million pages of printed books and pamphlets on medicine and related fields are to be digitised over two years.

Since the inception of the Wellcome's Digitisation Programme and its formal linking to the Trust's major research challenges, the Library's digital collections have been and are continuing to grow rapidly and to meet the organisation's goal to convert to digital form its holdings, whether they be books, videos and audio recordings, archives, manuscripts, paintings, prints, drawings, photographs, ephemera or yet other formats and media (Wellcome Library, 2015a).

3.14 ARCADIA FUND

In 2007, the Lisbet Rausing Charitable Fund, originally established in 2001, was renamed the Arcadia Fund. During its first seven years, it focused on preserving treasures of culture or nature, environmental conservation and human rights, and funding was disbursed to major projects at an international level. Since 2008, the Fund has concentrated exclusively on protecting endangered culture and nature to achieve lasting and greatest impact, and cultural grants have been available to institutions worldwide, including museums, archives and universities. The Fund's stated aim is 'to support the best projects in the field' (Arcadia Fund, 2007, p. 3).

Amongst the wide range of projects financed by Arcadia are several major digitisation initiatives. All are intended to preserve documentation and artefacts at risk, predominantly in or about the developing world. Although the digitised content relates largely to other countries' materials, two of the major projects, the *Endangered Archives Programme* and the *African Rock Art Image Project* have been supported by the British Library and British Museum, respectively, with full sets of the outputs being mounted on the UK institutions' websites in addition to holdings and access arrangements in the countries of origin. A further initiative, also with its output on the British Library website, is the *Oral History of British Science*.

The British Library hosts the content from the *Endangered Archives Programme*, which aims to preserve the archival material of 'pre-industrial' societies that is in danger of destruction, neglect or physical deterioration (*Arcadia Fund Annual Review*, 2007–2012; British Library, c2014). Reports available at the time of writing show that some 197 grants had been awarded in 70 countries (Arcadia Fund, 2012, p. 32). These cover initiatives about 'pre-modern' societies across Asia, Latin America, Africa and parts of Europe. A very broad definition of 'archives' is permitted, and it includes contents such as newspapers, periodicals and audio-visual resources. The full details of individual countries' initiatives can be found on the British Library website, as listed in Appendix 1. The *Oral History of British Science* recounts the events surrounding some of the most remarkable scientific and engineering discoveries of the past century, and the life and work of the scientists involved. The coverage is wide, and includes engineering, other physical sciences and biosciences.

The *African Rock Art Image Project* is a partnership between the British Museum and the Trust for African Rock Art (TARA). Africa's rock art is amongst the oldest and most varied of this art form, reflecting cultures that have long since disappeared. The Project is cataloguing and digitising TARA's photographic archive, to be available through the British Museum website in perpetuity (Arcadia Fund, 2012, p. 18).

3.15 COHERENCE OF CONTENT

The exemplar projects outlined in this chapter, and their outputs as listed in Appendix 1, are the result of not only individual initiatives, but also many cases of collaborative ventures. Often, however, such partnerships are short to medium term, and change as their strategic priorities alter, or different funding opportunities arise. Overall, the variety and disparity of the resources reflect the much regretted 'silos' discussed in Chapter 1. Viewed as part of the national resource, the content in the appendix appears to have very little coherence in terms of its selection. The following chapters will explore this issue further, by examining principal criteria for selection of intellectual content, how these criteria relate to the outputs in Appendix 1 and their potential for improving future planning of digitisation projects.

REFERENCES

Arcadia Fund. (2007) *Arcadia*. [Online] Available from: http://www.arcadiafund.org.uk/media/4983/2007.pdf.

Arcadia Fund. (2012) *Arcadia Annual Review 2012: protecting endangered nature and culture*. [Online] Available from: http://www.arcadiafund.org.uk/media/2354/2012.pdf.

Arcadia Fund Annual Review 2007-2012. [Online] Available from: http://www.arcadiafund.org.uk/annual-review.aspx.

Arts and Humanities Research Council. (c2011) *AHRC Theme Large Grants: call for outline proposals*. [Online] Available from: http://www.ahrc.ac.uk/Funding-opportunities/Documents/Theme%20Large%20Grants%20Call.pdf.

Becta. (2008) *Harnessing technology: next generation learning 2008-2014*. [Online] Available from: http://dera.ioe.http://webarchive.nationalarchives.gov.uk/20060315075935/http://dfes.gov.uk/publications/e-strategy/docs/e-strategy.pdfac.uk/8287/1/download.cfm%3FresID%3D37348.

Big Lottery Fund. (2006) Digitisation of Learning Materials and Community Grid for Learning: final evaluation findings. *Big Lottery Fund Research*, Issue 26. [Online] Available from:

http://www.google.co.uk/url?sa=t&rct=j&q=&esrc=s&source=web&cd=1&ved=0CCAQFjAA&url=http%3A%2F%2Fwww.biglotteryfund.org.uk%2Fglobal-content%2Fresearch%2Fwales%2Fdigitisation-of-learning-evaluation-findings&ei=8V61U9XoLeXG7AaNkoHQBQ&usg=AFQjCNFx2IdGWSRIqIkc2sydXIMYBwAvbg&sig2=MJFZgKiJ-cxi18vp-Uqnqg&bvm=bv.70138588,d.ZGU.

British Library. (1999-2015) *Annual Report and Annual Highlights* (1999-2000 to 2014-2015). [Online] Available from: http://www.bl.uk/aboutus/annrep/index.html.

British Library. (2008) *Digitisation Strategy 2008-2011.* [Online] Available from: http://www.bl.uk/aboutus/stratpolprog/digi/digitisation/digistrategy/.

British Library. (c2014) *The Endangered Archives Programme.* [Online] Available from: http://eap.bl.uk/pages/about.html.

British Library. (n.d.) *Digitisation.* [Online] Available from: http://www.bl.uk/aboutus/stratpolprog/digi/digitisation/.

Bruce, R. and Notay, B. (2004) The JISC 5/99 Programme: what's in a number?. *Ariadne.* Issue 38. [Online] Available from: http://www/ariadne.ac.uk/issue38/5-99.

Carr, R. (2007) *The Academic Research Library in a Decade of Change.* Oxford: Chandos. Information Professional Series.

Department for Education and Skills. (2005) *Harnessing technology: transforming learning and children's services.* Nottingham: DfES. [Online] Available from: http://webarchive.nationalarchives.gov.uk/20130401151715/https://www.education.gov.uk/publications/eOrderingDownload/1296-2005PDF-EN-01.pdf.

Dunning, A. (2010) Digitizing the past: next steps for public sector digitization. In Woodward, H. and Estelle, L (Eds.) *Digital Information: order or anarchy?* London: Facet, 117–131.

Dunning, A. (2011) *List of Projects Funded under UK New Opportunities Fund.* [Online] Available from: http://eprints.rclis.org/17518/.

Education for Change. (2006) *The Fund's ICT Content Programmes: final evaluation report.* London: Education for Change. [Online] Available from: http://www.biglotteryfund.org.uk/er_eval_ict_final_rep.pdf.

eLib. (2001) *The Projects.* Bath:UKOLN. [Online] Available from: http://www.ukoln.ac.uk/services/elib/projects/.

ESYS Consulting. (2000) *Summative Evaluation of Phases 1 and 2 of the eLib Initiative: Final Report.* Guildford: ESYS. [Online] Available from: http://opus.bath.ac.uk/35005/1/elib_fr_v1_2.pdf.

ESYS Consulting. (2001) *Summative Evaluation of Phase 3 of the eLib Initiative: Final Report.* Guildford: ESYS. [Online] Available from: http://opus.bath.ac.uk/35000/2/elib_eval_main.pdf.

Evidence Base. (2007) Evaluation of the JISC Digitisation Programme Phase 1 and International Contextualisation. Birmingham: UCE Birmingham. No current live URL.

Glenaffric. (2009) *Formative evaluation of the JISC Digitisation Programme Phase 2: final report.* JISC. [Online] Available from: http://www.jisc.ac.uk/media/documents/programmes/digitisation/digevalfinalreportf2_final_002.pdf.

Green, A. (1997) Towards the digital library: how relevant is eLib to practitioners? *New Review of Academic Librarianship,* Vol. 3, No. 1: 39–48.

Green, A. (2002) Digital Library, Open Library: developments in the National Library of Wales. *Alexandria.* Vol. 14, No. 3: 161–70. [Online] Available from: http://www.llgc.org.uk/fileadmin/documents/pdf/alexandria_S.pdf.

Griffiths, R. and Maron, N.L. (2009) *The National Archives (UK): Digitisation with Commercial Partnerships via the Licensed Internet Associates Programme.* (Ithaka Case Studies in Sustainability). [Online] Available from: http://sca.jiscinvolve.org/wp/files/2009/07/sca_bms_casestudy_natarchives.pdf.

Henshaw, C. (2009) Moving Image and Sound Collection. *Wellcome History*. No. 42: 20—21. [Online] Available from: http://www.wellcome.ac.uk/stellent/groups/corporatesite/ @msh_publishing_group/documents/web_document/wtx057796.pdf.

Henshaw, C. and Kiley, R. (2013) The Wellcome Library, Digital. *Ariadne*. Issue 71. [Online] Available from: http://www.ariadne.ac.uk/issue71/henshaw-kiley.

Henshaw, C., Savage-Jones, M. and Thompson, D. (2010) A Digital Library Feasibility Study. *Liber Quarterly*. Vol. 20, No. 1: 53—65. [Online] Available from: http://liber. library.uu.nl/index.php/lq/article/view/7975/8280.

House, D. (2012) An overview of e-resources in UK further and higher education. In Fieldhouse, M. and Marshall, A. (Eds) *Collection Development in the Digital Age*. London: Facet, 47—58.

House of Commons; Select Committee on Culture, Media and Sport. (2000) *Sixth Report*. [Online] Available from: http://www.parliament.the-stationery-office.co.uk/pa/cm 199900/cmselect/cmcumeds/241/24102.htm.

Jenkins, G. (2005) *Digitisation Policy and Strategy, National Library of Wales*. Aberystwyth: NLW.

Jisc. (1994) *JISC Circular 4/94: FIGIT Framework*. [Online] Available from: http://www.jisc. ac.uk/fundingopportunities/funding_calls/1994/10/circular_4_94.aspx

Jisc. (1995) *JISC Circular 11/95: Electronic Library Programme (eLib)*. [Online] Available from: http://www.jisc.ac.uk/fundingopportunities/funding_calls/1995/11/circular_11_95.aspx

Jisc. (1997) *JISC Circular 3/97: eLib Phase 3*. [Online] Available from: http://www.jisc.ac.uk/ fundingopportunities/funding_calls/1997/05/circular_3_97.aspx

Jisc. (1999) *Circular 5/99: Developing the DNER for Learning and Teaching*. [Online] Available from: http://www.jisc.ac.uk/fundingopportunities/funding_calls/2000/01/circular_5_ 99.aspx.

Jisc. (2006) *JISC Capital Programme: Digitisation: request for proposals*. JISC. [Online] Available from: http://www.jisc.ac.uk/media/documents/funding/2006/05/digitisation_rfp_ april_2006_final.pdf.

Jisc. (2008) *Digitisation Strategy*. [Online] Available from: [Online] Available from: https:// www.jisc.ac.uk/whatwedo/programmes/digitisation/jisc_digitisation_strategy_2008. doc.

Jisc. (2011) *Clustering and Sustaining Digital Resources: the JISC eContent Programme 2009-2011*. [Online] Available from: http://www.webarchive.org.uk/wayback/archive/201406 15013612/http://www.jisc.ac.uk/media/documents/publications/general/2011/JISCe ContentClusteringAndSustainingDigitalResources.pdf.

Jisc. (2013) *Jisc Strategy 2013-2016*. [Online] Available from: http://www.jisc.ac.uk/reports/ jisc-strategy-2013-16.

Jisc Digital Media. (n.d.) Developing Community Collections. [Online] Available from: http://www.jiscdigitalmedia.ac.uk/infokits/community_content/.

Joint Funding Councils' Libraries Review Group. (1993) *Report* [Follett Report]. Bristol: HEFCE.

Joint Funding Councils' Libraries Review Group. (1996) *Report* [Anderson Report]: report of the Group on a National/Regional Strategy for Library Provision for Researchers. [Online] Available from: http://www.ukoln.ac.uk/services/elib/papers/other/anderson.

Jones, R. A. (2008) A marathon not a sprint: lessons learnt from the first decade of digitisation at the National Library of Wales. *Program*. Vol. 42, No. 2: 97—114.

Llewellyn, M. (2012) *Current Future Strategies: Digital Transformations in Arts and Humanities*. Paper presented at the AHRC Digital Research Resources in the Workshop, 25 July.

Mahoney, J. (1998) Introduction. In Carpenter, L., Shaw, S. and Prescott, A. (eds) (1998) *Towards the Digital Library: the British Library's Initiatives for Access Programme*. London: British Library. 10—14.

Maron, N.L. (2011) *The National Archives (UK): Enhancing the Value of Content through Selection and Curation: Case Study Update.* Ithaka Case Studies in Sustainability. [Online] Available from: http://sr.ithaka.org/sites/default/files/reports/SCA_IthakaSR_CaseStudies_TNA_2011.pdf.

McMenemy, D. (2009) *The Public Library.* London: Facet.

Milne, R. (2002) The Distributed National Collection Access, and Cross-sectoral Collaboration: the Research Support Libraries Programme. *Ariadne.* Issue 31. [Online] Available from: http://www.ariadne.ac.uk/issue31/rslp.

Mowat, I. R. M. (1998) The Non-Formula Funding of Special Collections in the Humanities initiative. *Library Review.* Vol. 47, No. 5: 301–305.

National Archives of Scotland. (2000) *Annual Report of the Keeper of the Records of Scotland 1999-2000.* [Online] Available from: http://www.nas.gov.uk/about/annualReport.asp.

National Archives of Scotland. (2001) *Annual Report of the Keeper of the Records of Scotland 2000-2001.* [Online] Available from: http://www.nas.gov.uk/about/annualReport.asp.

National Archives of Scotland. (2003) *Annual Report of the Keeper of the Records of Scotland 2002-2003.* [Online] Available from: http://www.nas.gov.uk/about/annualReport.asp.

National Archives of Scotland. (2004) *Annual Report of the Keeper of the Records of Scotland 2003-2004.* [Online] Available from: http://www.nas.gov.uk/about/annualReport.asp.

National Archives of Scotland. (2005) *Annual Report of the Keeper of the Records of Scotland 2004-2005.* [Online] Available from: http://www.nas.gov.uk/about/annualReport.asp.

National Archives of Scotland. (2006) *Annual Report of the Keeper of the Records of Scotland 2005-2006.* [Online] Available from: http://www.nas.gov.uk/about/annualReport.asp.

National Archives of Scotland. (2007) *Annual Report of the Keeper of the Records of Scotland 2006-2007.* [Online] Available from: http://www.nas.gov.uk/about/annualReport.asp.

National Archives of Scotland. (2010) *Corporate Plan 2010-2011 to 2012-2013.* [Online] Available from: http://www.nas.gov.uk/documents/Corporateplan2010-2013.pdf.

National Archives of Scotland. (2011) *Annual Report of the Keeper of the Records of Scotland 2010-2011.* [Online] Available from: http://www.nas.gov.uk/about/annualReport.asp.

National Audit Office. (2004) *The British Library: providing services beyond the Reading Rooms.* London: Stationery Office. HC 879 (2003-2004). [Online] Available from: http://www.nao.org.uk/wp-content/uploads/2004/07/0304879.pdf.

National Library of Scotland. (1997-2014) *Annual Review and Annual Report.* [Online] Available from: http://www.nls.uk/about-us/publications/annual-review.

National Library of Scotland. (2005) *Digital National Library of Scotland: Strategic Plan 2005-2008.* [Online] Available from: http://www.nls.uk/media/22406/nls_digital_library_strategy.pdf.

National Library of Scotland. (2008) *Integrated Collecting Strategy.* [Online] Available from: http://www.nls.uk/media/22389/2008-collecting-strategy.pdf.

National Library of Wales. (2009) *Digitisation Strategy 2008/9 – 2010/11.* [Online] Available from: http://www.llgc.org.uk/fileadmin/fileadmin/docs_gwefan/amdanom_ni/dogfennaeth_gorfforaethol/dog_gorff_strat_dig_08_09_10_11S.pdf.

National Library of Wales. (2011) *The Theatre of Memory: Welsh Print Online.* [Aberystwyth]: National Library of Wales. [Online] Available from: http://www.llgc.org.uk/fileadmin/fileadmin/docs_gwefan/amdanom_ni/dogfennaeth_gorfforaethol/dog_gorff_dog_thycS.pdf

National Library of Wales. (2012). *Annual Review 2011/12.* [Aberystwyth: National Library of Wales]. Online at http://www.llgc.org.uk/fileadmin/fileadmin/docs_gwefan/amdanom_ni/dogfennaeth_gorfforaethol/dog_gorff_abl_11_12S.pdf.

Parkinson, N. and Bruce, R. (comps) (c2000) *Accessing our Humanities Collections: a subject guide for researchers.* London: Jisc.

Pearson, D. (2001) Developing the Wellcome 'Digital Library". *Wellcome History*. Vol. 18, No. 9. [Online] Available from: http://www.wellcome.ac.uk/stellent/groups/corporatesite/@msh_publishing_group/documents/web_document/wtd006085.pdf.

Pinfield, S. (2004) eLib in retrospect: a national strategy for digital library development in the 1990s. In Andrews, J. and Law, D. (eds) *Digital Libraries: policy, planning and practice*. Aldershot: Ashgate, 19–34.

Pinfield, S. and Dempsey, L. (2001) The Distributed National Electronic Resource and the hybrid library. *Ariadne*. Issue 26. [Online] Available from: http://www.ariadne.ac.uk/dner.

Poole, N. (2009) *How to save NOF Digi?* [Online] Available from: http://www.collectionstrust.org.uk/how-to-save-nof-digi/.

Public Record Office of Northern Ireland. (2006-2014) *Deputy Keeper's Report*. [Online] Available from: http://www.proni.gov.uk/index/about_proni/who_are_we_and_what_we_do/business_plans_deputy_keepers_reports_and_annual_reports.htm.

Rikowski, R. (2011) Digitisation: research, sophisticated search engines, evaluation – all that and more. In Rikowski, R. (Ed.) *Digitisation Perspectives*. Rotterdam: Sense, 65–86.

Robey, D. (2008) *Sustainability of AHRC-funded Digital Resources*. [Online] Available from: http://www.europaeum.org/files/publications/sustainability%20issues.pdf.

RSLP: Research Support Libraries Programme. (2002) [Online] Available from: http://www.rslp.ac.uk/AboutUs/.

Rusbridge, C. (1998) Towards the Hybrid Library. *D-Lib Magazine*. July/August. [Online] Available from: http://www.dlib.org/dlib/july98/rusbridge/07rusbridge.html.

Rusbridge, C. (2001) After eLib. *Ariadne*. Issue 26. [Online] Available from: http://www.ariadne.ac.uk/issue26/chris/.

Shenton, H. (2009) Virtual reunification, virtual preservation and enhanced conservation. *Alexandria*. Vol. 21, No. 2, 33–45.

Sykes, J. (2008) Large-scale digitisation: the £22million JISC programme and the role of libraries. *Serials*, Vol. 21, No 3: 167–173.

Tedd, L. (2011) 'People's Collection Wales'. *Program*, Vol. 45, No. 3: 333–345.

The National Archives. (2005) *The National Archives Digitisation Programme 2005-2011*. [Online] Available from: http://www.nationalarchives.gov.uk/documents/digitisation-programme2005-2011.pdf.

The National Archives. (2007) *Strategic Plan 2007 to 2008*. [Online] Available from: http://collections.europarchive.org/tna/20110131174836/http://www.nationalarchives.gov.uk/documents/strategic-plan0708.pdf.

The National Archives. (2009) *The National Archives Digitisation Programme 2009-2013*. [Online] Available from: http://collections.europarchive.org/tna/20110131174836/http://www.nationalarchives.gov.uk/documents/digitisation-programme-2009-2013-revised-march-2010.pdf.

The National Archives. (2012) *Archives for the 21st Century in action: refreshed 2012-2015*. [Online] Available from: http://www.nationalarchives.gov.uk/documents/archives/archives21centuryrefreshed-final.pdf.

The National Archives. (2015a) *Catalogues and Online Records*. [Online] Available from: http://www.nationalarchives.gov.uk/records/catalogues-and-online-records.htm.

The National Archives. (2015b) *Digitisation and Digital Archives*. [Online] Available from: http://www.nationalarchives.gov.uk/about/websites-digitisation-digital-archives.htm.

The National Archives. (2015c) *Licensing our Records*. [Online] Available from: http://www.nationalarchives.gov.uk/commercial/licensing.htm.

The National Archives. (2015d) *Our Online Records*. [Online] Available from: http://www.nationalarchives.gov.uk/records/our-online-records.htm.

Wellcome Library. (2015a) *Digitisation at the Wellcome Library.* [Online] Available from: http://wellcomelibrary.org/what-we-do/digitisation/.

Wellcome Library. (2015b) *Digitisation Schedules.* [Online] Available from: http://wellcome library.org/what-we-do/digitisation/digitisation-schedules/.

Wellcome Library. (2015c) *Transforming the Wellcome Library: 2009-2014.* [Online] Available from: http://wellcomelibrary.org/what-we-do/library-strategy-and-policy/transforming-the-wellcome-library/.

Wellcome Trust. (2003-2013) *Annual Report and Financial Statements* [2003-2013]. [Online] Available from: http://www.wellcome.ac.uk/About-us/Publications/Annual-Report-and-Financial-Statements/Previous/index.htm.

Wellcome Trust. (2006) *Free Online Access to nearly Two Hundred Years of Medical Research.* [Online] Available from: http://www.wellcome.ac.uk/News/Media-office/Press-release-archive/WTX031289.htm.

Wellcome Trust. (2009) A moving sight. *Wellcome News.* Vol. 61, No. 9. [Online] Available from: http://www.wellcome.ac.uk/stellent/groups/corporatesite/@msh_publishing_group/documents/web_document/WTX057848.pdf.

Wellcome Trust. (2010) *Wellcome Library launches major digitisation project.* [Online] Available from: http://www.wellcome.ac.uk/News/2010/News/WTX062533.htm.

Wellcome Trust. (2012) *Research Resources in Medical History: Directory of Grants awarded 2000/01 to 2011/12.* [Online] Available from: http://www.wellcome.ac.uk/stellent/groups/corporatesite/@policy_communications/documents/web_document/wtvm056353.pdf.

Wellcome Trust. (2013a) *London's Pulse opens up the capital's health records.* [Online] Available from: http://www.wellcome.ac.uk/News/Media-office/Press-releases/2013/Press-releases/WTP054580.htm.

Wellcome Trust. (2013b) *Million-page story of modern genetics launched by the Wellcome Library.* [Online] Available from: http://www.wellcome.ac.uk/News/Media-office/Press-releases/2013/WTP051905.htm.

Wellcome Trust. (2014a) *Wellcome Trust Library Funds a New Partnership to Digitise 800,000 Pages of Mental Health Archives.* [Online] Available from: http://www.wellcome.ac.uk/News/Media-office/Press-releases/2014/WTP057722.htm.

Wellcome Trust. (2014b) *Wellcome Trust and NLM establish agreement to make 150 years of biomedical journals freely available online.* [Online] Available from: http://www.wellcome.ac.uk/News/Media-office/Press-releases/2014/WTP056252.htm.

Wellcome Trust. (2014c) *Jisc and Wellcome partnership to create comprehensive online resource for the history of medicine.* [Online] Available from: http://www.wellcome.ac.uk/News/Media-office/Press-releases/2014/WTP055999.htm.

Wellcome Trust. (2014d) *Wellcome Library and Jisc announce partners in 19th-century medical books digitisation project.* [Online] Available from: http://www.wellcome.ac.uk/News/Media-office/Press-releases/2014/WTP056966.htm.

Winterbottom, A. and Robey, D. (c2010) *Report on AHRC-funded Digital Research Resources.* [Online] Available from: http://www.arts-humanities.net/files/ahrc_data_resources_report.pdf.

Woodhouse, S. (2001) The People's Network and the Learning Revolution: building the NOF Digitize Programme. *Ariadne.* Issue 29. [Online] Available from: http://www.ariadne.ac.uk/print/issue29/woodhouse.

CHAPTER 4

Content Selection for Digitisation: Principal Criteria and Mapping of UK Outputs

The disparate nature of the strategic environment described in Chapters 1 and 2, combined with the variety of apparently unconnected content outlined in Chapter 3 and Appendix 1, turn the spotlight on how the selection of intellectual content for digitisation could be coordinated more effectively to improve the coherence of content in the totality of national provision. It is an awkward question, because, unlike other elements of digitisation work, selection criteria have not been the subject of efforts to standardise, and any such exercise runs the risk of being too narrow and therefore too constraining when selecting against the reality of particular stakeholders' interests. Nonetheless, effective selection is at the heart of all digitisation work, and it is valuable to consider what progress could be made in its coordination.

Firstly, however, it is necessary to ask whether this issue is still relevant with the advent of mass digitisation. In the last decade, this step change in digital provision has raised hopes that the complexities of a content selection process may soon be irrelevant when planning digitisation and that comprehensiveness can become a realistic aim. Two principal trends serve this purpose. The first is the international initiatives to gather digital content in mass quantities for access via the Internet. Those of most recent years include *Google Books*,[1] *Microsoft Live Book Search* (content from which ultimately became available through the *Internet Archive*),[2] the *Open Content Alliance*[3] and the *Hathi Trust*,[4] all standing alongside longer established services such as *JSTOR*[5] and *Project Gutenberg*,[6] the earliest of them all. Together, they have greatly advanced the quantity of digitised content now

---

[1] https://books.google.co.uk.
[2] https://archive.org/index.php.
[3] http://www.opencontentalliance.org.
[4] https://www.hathitrust.org.
[5] http://www.jstor.org.
[6] https://www.gutenberg.org.

Stepping Away from the Silos
ISBN 978-0-08-100278-0
http://dx.doi.org/10.1016/B978-0-08-100278-0.00004-0

available online, and are developing the knowledge and expertise to address the many factors and issues that are specific to this part of the digital landscape. The second trend is the development of major aggregators, such as *Europeana*,[7] the *Digital Public Library of America (DPLA)*[8] and *DigitalNZ*,[9] which furnish mass provision of distributed resources by drawing together the outputs of multiple contributors into one readily accessible service.

The experience of these initiatives to date does show promise that, over time, mass provision could indeed bring much if not all of our analogue content into the digital environment. It has also become clear, however, that there are major constraints which delay progress, such as Google's issues with copyright restrictions, quality of outputs, and rights of access to full texts. It is also true that even in these initiatives, selection criteria, albeit at a very broad level, do still apply. To date, books predominate in several of the initiatives, and *JSTOR*'s principal focus for many years has been as a leading provider of journal literature. In the case of aggregators such as *Europeana, DPLA* and *DigitalNZ*, the selection is strongly influenced not only by the holdings and decisions of the originating libraries, but also by their geographic location.

Viewed as selection criteria, however, these are very wide indeed, and they do not detract from the potential to build comprehensive provision over time. Even if this is a realistic prospect, however, such a day is still distant. Local, national and global knowledge and heritage are represented in over 500 years of printed materials; in manuscripts, images and artefacts from antiquity onwards; and in the audio and video technologies of the 19th to 20th centuries. The sheer quantity of analogue content still requiring conversion to digital form is potentially overwhelming, both in terms of workload and of financial support. The expectation of and demand for digital versions of these materials will not abate. Digitisation work short of mass digitisation will continue, and selection criteria will be applied, whether in a structured or unstructured fashion.

The content delivered by the major projects described in Chapter 3 and Appendix 1 as exemplars of UK digitisation work illustrate many variations in approach to intellectual content selection. In terms of scope and coverage, the projects in Appendix 1 are very wide-ranging, from large-scale initiatives providing major interdisciplinary resources to small and

---

[7] http://www.europeana.eu/portal/.
[8] http://dp.la.
[9] http://www.digitalnz.org.

highly specific materials (in some cases, individual items). This reflects the variant aims and purposes of the projects as outlined in Chapter 3. It also indicates marked variations in the selection criteria used by the different projects to determine the intellectual content to be digitised. This latter issue results not only from the different strategic contexts in which the initiatives were conceived, but also from a looser approach amongst practitioners to setting such criteria. Compared to the more concentrated and structured efforts made since the 1990s to establish standards for other aspects of digitisation, the attention to standardising selection criteria has been more limited. It is not unusual to hear it said that each project probably has unique selection criteria, and systematic studies have come to similar conclusions. A detailed review in 2009 of international practice and theoretical literature on the digitisation of heritage collections observed that 'when they are not explicitly stated to follow existing guidelines, selection practices appear most often based on ad hoc decisions or on available funds. The criteria underlying selection seem as numerous as the digitisation projects themselves' (Ooghe and Moreels, 2009). Daigle perceives such matters as undermining the very heart of digital provision, when commenting that 'in many ways, the most important part of digital curation is the selection and creation of digital content. This is where most organizations struggle with their overall stewardship strategy and it requires special attention' (Daigle, 2012, p. 95).

In fact, core objective guidelines and criteria for the selection of digital content have featured within the wider work on digitisation standards over the last twenty years, albeit less prominently than the other aspects of digitisation, and common elements can be identified from the wider literature. These not only supply a means of more consistent planning of new digitisation work, but could play a role in identifying underlying coherence in existing, disparate outputs such as those covered in Chapter 3 and Appendix 1.

4.1 CONTENT SELECTION CRITERIA: DEVELOPMENT OF PRINCIPAL COMMON CRITERIA

The development of objective criteria for content selection and their consistent application has undoubtedly been erratic over the 20 years of digitisation activity since the 1990s. It is common to find that the original selection decisions for many digitisation initiatives cannot be traced, or, indeed, have never existed in any formal statement. Where they have been devised, they have often been withheld from the public domain. When they have been available, the criteria have been characterised by disparate approaches and vocabulary, creating in turn practices best described as 'trial and error' (Ooghe and Moreels, 2009). Gertz has taken the view that 'no absolute criteria guide selection for digitization, only questions to be addressed within the context of the individual institution' (Northeast Document Conservation Center, c2007; Gertz, 2013). In other literature about initiatives in libraries and closely related fields, there has been much output on the planning and running of digitisation initiatives, but a considerable amount of the documentation on selection has focused principally or totally on IT and legal matters, with scant reference if any at all to the intellectual issues about content. This is not as unreasonable as it may seem at first sight. As noted in Chapter 1, the advent of digitisation required that many entirely new elements in the technical, infrastructural, discovery, preservation, legal and financial areas had to be developed, consolidated and matured. These were complex fields that changed rapidly. It is understandable that there was much emphasis on the creation of robust standards and coordination of such practice; they were imperatives to apply the new technology effectively. As a consequence, guidance and criteria for the selection of content in terms of its intellectual value, a field that was already very familiar to information and heritage practitioners in relation to analogue materials, was somewhat overshadowed.

This was probably not, however, the only factor at play, at least in the earliest years of digitisation. The challenge of prioritising when choosing from the vast analogue base of information and heritage materials cannot be underestimated. It was a dilemma that was recognised in the very early days, as reported by Chris Rusbridge in relation to the eLib Programme. 'In digitisation we suffered from the indecision of the pioneer; although anything we did was valuable, nothing appeared of itself likely to justify the very high cost of digitisation. This was an area where consistent belief in a vision of the increasing value of contributions towards a critical mass of digitised material is needed' (Rusbridge, 2001).

Nonetheless, over time project work did address the intellectual aspects of content selection issues. This has not produced standards for the selection of intellectual content, but rather an identifiable corpus of guidelines, that has developed gradually out of individual programmes and initiatives from the late 1990s onwards. Appendix 2 lists the relevant documentation that has been consulted for this chapter. Several of the titles offer advice that is independent of any one organisation's practice. Others are influential guidelines created originally within leading institutions for their own purposes, but subsequently adopted by the wider community. In addition the appendix includes references to websites that list fuller bibliographies on both content issues and the wider aspects of selection for digitisation.

The exact coverage of these guidelines varies. Certain focus on cultural heritage materials in their widest sense (NINCH, 2003; NISO, 2007; Federal Agencies Digitization Guidelines Initiative, 2009) while the others concentrate on documentary collections, in most cases very broadly defined to include materials such as printed works, manuscripts, archives, images, maps, and audiovisual items. Not unconnected to this is the fact that many of these outputs originate from the library sector. In their survey, Ooghe and Moreels found 'a clear bias towards specific documentary types or sectors, with libraries being well represented, but museums and art galleries offering particularly limited basis for their review of practice' (Ooghe and Moreels, 2009).

It is noteworthy that the titles in Appendix 2 are separate publications from those on overall collection development and management in heritage organisations. With technology now being all pervasive, one might assume that collection development practices in heritage institutions such as libraries, archives and museums would have integrated professional litera-ture and practice for the selection of all materials, whether they be analogue, born digital or digitised. Since the 1990s, the reality has been more mixed. At the theoretical level there have been differences of opinion. As early as 1997, Johnson placed strong emphasis on the importance of selection processes that are blind to format and delivery mechanisms (Johnson, 1997, p. 88). By contrast, Ooghe and Moreels argue that in theory there is a conceptual difference between selection for digitisation and selection for collection management, although they do acknowledge that in practice their survey evidence found significant overlap between the two exercises (Ooghe and Moreels, 2009). It is notable that in the library-specific literature many of the practices guiding content selection for digitisation have been articulated independently of guidance on wider collection issues. Many books and documents in the field do not cover selection for digitisation in any significant detail or give scant reference to this aspect. Others concentrate on the overall concept of digital library development, with the emphasis on electronic materials that are purchased, licensed or available on open access through the Internet, and give only limited coverage of selection for digitisation (for example, Johnson, 1997; Lesk, 2005). By contrast, however, at the level of individual institutions' policies, one is more likely to find criteria for digitisation that are linked into the organisation's main collection development policies. Examples from Appendix 2 include the documentation from Yale and the Digital Library of Georgia (Yale University Library, 2006; Digital Library of Georgia, 2001).

In the core literature identified on content selection criteria, there have been several stages of development. There is early activity to devise initial general principles and working methods, and this covered key elements in varying ways. The much-referenced Harvard matrix developed as part of that university library's leading initiatives in the 1990s (Hazan et al., 1998). Other significant guidance in this early period included work by the Library of Congress (Library of Congress, n.d), the University of Oxford (Lee, 1999) and a key paper for the *Joint RLG and NPO Preservation Conference* in 1998 (Ayris, 1998). From 2000 onwards a period of consolidation followed, which brought extensive and, in some cases, comprehensive documentation on all elements of digitisation. Within these, however, due attention and weight were given to the intellectual aspects of content selection (IFLA, 2002; NINCH, 2003; Cornell University Library, 2004; NEDCC, c2007; NISO, 2007; Federal Agencies Digitization Guidelines Initiative, 2009; Ooghe and Moreels, 2009; National Library of Australia, c2012; Gertz, 2013; UNESCO, n.d.). Later guidelines concentrate predominantly on selection of intellectual content (Yale University Library, 2006; DigitalNZ, 2009; Jisc Digital Media, 2015; University of California Libraries, 2013).

In isolating from these documents the common criteria for selection of intellectual content, it becomes clear that, as with all other aspects of digitisation, this is a complex area. The principal criteria that emerge are value, uniqueness and rarity, added value, user needs, thematic and subject coverage, format and medium, coherence, virtual reunification and clustering. There is a natural tendency in the literature to consider each of these as an independent entity, and this is reinforced by legitimate advice that not all criteria will apply to all initiatives. There are, however, inter-relationships between the different criteria which should be understood and accommodated when planning for digitisation, and this chapter considers the linkages between the principal criteria, as well as their significance as individual factors in content selection.

Before examining the criteria in detail, it is important to note the wider context in which they are set. The literature consistently stresses that content selection should be made to meet the strategy of the parent organisation or institution. There is a paucity of attention to the potential conflicts of strategic interest that could arise in collaborative projects, despite the emphasis on collaborative ventures that is found in many countries, and certainly in the UK, as observed in Chapters 2 and 3. While the criteria are indeed relevant to such circumstances, it is important to remember that their application could be subject to considerable modification to accommodate a collaborative initiative.

Much, though not all, of the work on content selection focuses specifically on digitisation to support research. In practice, however, the criteria discussed later can apply, with careful tailoring, to digitisation for different uses and constituencies, from the highly specialised to the widest of interests.

Whatever the target community, it is important to set clear aims and purpose for the content to be selected. These can vary quite widely depending on the strategic context, but perhaps the most universal issue is the choice between digitisation for access or for preservation. There are many definitions for these two terms. For the purposes of the present discussion, 'access' is taken to mean 'both broader access to a global audience via the internet, and enhanced access, by making collections searchable, findable and linked to related materials' (Hughes, 2012, p. 6). 'Preservation' is accepted as the '[maintenance] of long-term access to the objects being preserved or to the information that they contain. Preservation usually involves making a copy of original material to 'back up' or to reduce wear on a non-digital item...On some occasions it may focus on last-chance migration of content from deteriorating carriers' (DigitalNZ, 2009).

In the earlier years of digitisation, there was a tendency to seek a strict choice between one or the other, and the priority between these two functions has been the source of much discussion and argument. As Shenton has observed, 'there is a continuing debate as to whether digitization is a preservation medium per se or whether it is an access medium with preservation dividends' (Shenton, 2009, p. 38). The exchanges have had many twists. Some of the earliest core documents took very clear positions on this matter, for example at the University of Oxford, where librarians and curators considered increased access to be the main benefit of digitisation (Lee, 1999), while over time there were major changes of viewpoint by other experts in the field. For example, Erway's earlier advocacy of preservation shifted to access for special collections as technological developments changed the environment (Erway, 2008). There have also been approaches that reconcile these differences and view the functions as complementary. 'Digitisation debates and policies often pose access and preservation as two competing priorities in tension with each other, or place preservation as a sub-set of access. By instead viewing access and preservation strategies as ways to address different points in a continuum of time, digitisation can be seen as a technology solution that can be applied according to the situation' (DigitalNZ, 2009). For the purposes of this book, this last view is adopted.

4.2 SELECTION CRITERIA FOR INTELLECTUAL CONTENT IN DIGITISATION PROGRAMMES

It is recognised that in practice the application of content selection criteria does not happen in isolation. When initiating a digitisation programme, technical, infrastructural, legal and preservation criteria will also apply and in some cases will take priority over preferred aspects of the selection issues. Arguably, however, it is the actual content that is central to digitisation programmes, and therefore, normally, its selection should be carefully assessed against criteria such as those discussed *before* the other aspects of the work are considered in detail.

The following sections discuss in more detail the nature of each of the principal content selection criteria identified in the core literature, and consider how they are reflected in the resources described in Chapter 3 and presented in Appendix 1. In the absence of original criteria for many of the existing outputs, a broad mapping of each project's content has been established against two or more of the selection criteria in the following sections, based on information from the project websites. This offers an indicative picture of the types of content delivered overall, and of areas of actual or potential critical mass that can be identified from the outputs examined. The figures on which percentages are based are presented in Appendix 3.

4.3 VALUE

Common to all content selection is the concept of value. 'Value' in the context of digitisation can be defined in many ways, and it inevitably carries certain levels of subjectivity (NINCH, 2003). Gertz advises that 'specific definitions of value and importance vary, but they cluster around intellectual, historic, and physical characteristics' (Gertz, 2013, p. 4). A more expanded version of these is offered by Ooghe and Moreels, who speak of 'such aspects as socio-historical, cultural, aesthetic or scientific meaning, production processes, public interest, formal language or technology' (Ooghe and Moreels, 2009). Whatever the initial definition chosen when selecting for a digitisation project it is important to reassess it regularly, as the full value of a digitised resource can take time to evolve and be understood (Hughes, 2012, pp. 1, 10).

4.3.1 Value: Uniqueness and Rarity

A wide range of factors can define the value of digitised content, often driven by the strategy and collection policies of the initiating organisations. This is very evident from the scope and variety of content described in Chapter 3 and Appendix 1. Within this range, however, two selection criteria, uniqueness and rarity, are especially prominent, and appear to transcend the boundaries of individual organisations' interests. Worldwide, these factors carry a dominant role, and in the UK their importance became evident from the earliest projects. In the 1990s many of the early adopters included such materials in their first ventures, such as the British Library's Initiatives for Access programme which produced digital versions of the *Lindisfarne Gospels*, the *Magna Carta* and the *Notebook of Leonardo da Vinci*. The postproject report on the e-Lib programme established that uniqueness and rarity were considered very important by participants as selection criteria (ESYS Consulting, 2000). This thinking has consolidated and strengthened both in the UK and beyond as digitisation work has matured. In 2008, Erway endorsed the view of Richard Ovenden, then Keeper of Special Collections at the Bodleian Library, University of Oxford, that special collections, being by definition unique or rare materials, are distinguishing features for institutions, and are therefore priority candidates for digitisation (Erway, 2008, p. 325). Coming up to the present day, Calhoun stresses that digitisation of special collections is becoming increasingly important and refers to a number of reports that recommend raising the priority of such holdings to enable their online discovery and use (Calhoun, 2014, pp. 118–9).

The outputs in Appendix 1 demonstrate that uniqueness and rarity have consistently been one of the leading criteria in UK initiatives since the 1990s. As can be seen from scanning the lists, these cover not only iconic items, often referred to as 'Treasures', but works of intellectual importance that go beyond the aesthetically beautiful or historically significant. The broad mapping mentioned previously indicates that at least 60% of the projects in Appendix 1 focus on unique materials only. (This estimate excludes unique items that have been combined with rare and other content to deliver a broader output.) Of these materials, the vast majority are small-scale initiatives delivering specialised content. There are also, however, some significant programmes that, over time, have come to deliver an important level of critical mass, for example the projects creating *British Library Sound* (originally the *British Library Sound Archive*); the *Census* and related materials from The National Archives, the National Records of Scotland and the Public Record Office of Northern Ireland; and the *British Pathe Film Archive*, digitised under the New Opportunities Fund.

4.3.2 Value: Added Value

If selecting content for its intrinsic value, however defined, is a central element in digitisation planning, this factor has an important extension in the potential to enhance the original content. 'The argument in favour of digitization is especially strong when there is an opportunity to add value' (Gertz, 2013, p. 7). 'Added value' is by no means a new element in digitisation work. By 1998, the Harvard guidelines had already identified that 'ideally, the electronic version of a source will permit new kinds of use and more sophisticated types of analysis' (Hazan et al., 1998). Around the same time, Ayris wrote that use of the content and links to embedded resources should be covered by digitisation projects, as well as use of standard software tools for studying the digitised resources (Ayris, 1998). From the earliest initiatives in the UK and beyond, deliverables have included not only improved search and manipulation functions, but contextual and interpretive information, links to related resources, learning packages, text and data mining facilities and many other types of enhancement, all becoming more sophisticated as experience and technology have progressed. In recent years, there has been an increasing emphasis on this aspect of digitisation, because 'users now expect more than rich collections' (Calhoun, 2014, p. 212).

4.3.3 Value: User Needs

User interests, needs and demand are central factors in defining value. The importance of making decisions to digitise based on user needs and demand is long and well acknowledged. Gertz argues that they are the vital second factor in the selection process, coming directly after value (Gertz, 2013). To some readers this may seem so obvious as to be a given, but actual practice in digitisation over the last 20 years has not matched the theory. There are many studies that have discovered limited or no research into user needs for digitised materials in the UK and beyond. A major report on digitised content in the UK observed that 'there is no overview of user needs and demand for digitised content in the UK' (Bültmann et al., 2005, p. 4). In a 2008 survey by Jisc of Head and Senior Librarians in HE and FE, only 41% of HEIs reported that they included some form of user consultation to prioritise collections for digitisation (Marchionni, 2009). In more recent years, assessments of practice in these fields remain somewhat negative. In 2013 Maron and Pickle observed that 'investments in understanding the needs of the audience [for a digitised collection] are quite low' (Maron and Pickle, 2013, p. 2). In the specific field of Digital Humanities, there is evidence that many resources for this discipline 'are still designed without reference to user requirements' (Warwick et al., 2012; xv). Dobreva et al. are yet more forceful in their assessment. 'In fact, we are currently witnessing a paradox: major institutions from the cultural heritage sector clearly emphasize the place of user evaluation and feedback in digitization-related policies. But in reality, decisions about aspects of digitization that impact users are frequently taken without direct user involvement' (Dobreva et al., 2012, p. 73).

Where information about user interests, needs and demand is lacking, a typical default is selection of content by individuals or teams involved in the management of the analogue materials. Undoubtedly, this brings the advantages of detailed knowledge of collections and informed vision about realising their potential. Nonetheless, it has been found that, without complementary information about user needs, there are real risks in 'push' initiatives. Calhoun warns that creating digitised resources based on 'build it and they will come' carries a high possibility of failure (Calhoun, 2014, p. 165); or as McMenemy expresses it, 'you could choose material that is particularly liked by you or other staff members rather than material that would be popular with customers' (McMenemy, 2005, p. 166). Such predictions have been borne out in various studies of use, such as the log analysis in 2008 which showed that one-third of digital resources in the humanities remained unused (Warwick, 2012, p. 5).

If many ventures have found it problematic to plan on the basis of user needs, this is probably because it is an area that is as complex in digitisation as it is in other aspects of educational and cultural activities. 'Digital resources are valuable to different audiences for different reasons, and some value may not be realized immediately' (Hughes, 2012, p. 5). The field is further complicated by the fact that needs and audiences can shift over time (Calhoun, 2014, p. 167). Target audiences can be researchers, teachers, students, specialist communities, or the general public, as well as innumerable subsets of these principal groups. In some instances, initiatives may quite legitimately address several such groups. A recurring but typical example of the difficulties arises when targeting a resource at user groups in both the educational and wider public sectors. 'The trade-off between popular content and content that might be deemed to be educationally important is an important one to strike' (McMenemy, 2005, p. 167). It can, however, be worth the effort to address multiple audiences. As Tanner says, 'widespread access to digitized resources contributes to the vibrant cultural and intellectual life of the nation, promoting education and enjoyment for all while bestowing a range of benefits to local and national economies' (Tanner, 2011, p. 115).

Despite the aforementioned complexities, there has been progress in the application of information about user needs to digital developments. Calhoun cites several examples of evidence that user consultation in many forms has contributed to successful digital library programmes (Calhoun, 2014, pp. 123, 166 and 170). While information about this is incomplete in relation to the projects in Appendix 1, there have been conscious efforts in the UK to apply user needs to digitisation choices. For example, the National Library of Wales has included user demand and needs in its digitisation strategies (NLW, 2009, 2011). The DiSCmap project investigated in some detail users' needs for further digitisation of special collections (Birrell et al., 2009). The Jisc Digitisation Programme Phase 2 (2007—09) placed a strong emphasis on collections that provided clear evidence of user needs having been identified (Marchionni, 2009), and the extensive range of guides from the Strategic Content Alliance (SCA) on audience development and engagement has relevance to digitisation as well as the other aspects of digital content provision that they also serve (Strategic Content Alliance, 2012).

As activity in this field has become more consistent in digitisation work, a range of methods has evolved and been recognised over time, all of which do feature in some of the projects in Appendix 1. These include analysis of usage statistics (see for example IFLA, 2002; NINCH, 2003; Erway, 2008); steering groups of key experts to advise on selection (see for example Marchionni, 2009); and user groups who inform selection and ways of using the information and give feedback on usability of the outputs. Dobreva et al. present a fuller analysis and discussion on these and related methods (Dobreva et al., 2012). All such methods can be applied in numerous permutations as suited to individual projects, and generally are best employed as part of the entire lifecycle of the project, that is before, during and after creation (Marchionni, 2009).

These methods for identifying user needs to support selection decisions are not in themselves new. In one form or another they will be familiar to practitioners in all heritage sectors as tools for managing and developing all aspects of their services. With the advent of social media, a genuinely new means of developing information resources is developing, in the form of crowdsourcing. To date, it appears that digitisation initiatives have embraced this principally to draw in the skills of communities or individuals to transcribe, correct and index a predetermined collection of materials. Initiatives in the UK include the *Old Weather Project* (Shuttleworth, 2015), and the *East London Theatre Archive* (Jisc, 2011).

There is, however, a role for crowdsourcing as a method for content selection. As far as it has been possible to establish, this has yet to appear in core literature on criteria for selection of digitised content, but some projects have explored this aspect. A prime example in the UK is the *Great War Archive* (Jisc, 2010). This site was established as part of the Jisc-funded *World War I Poetry Digital Archive*. It invited the general public to contribute items in private ownership. Responses were considerably higher than expected, and some 6500 additional items were brought into the public domain, including personal letters and diaries, recordings of interviews with veterans, physical artefacts and much more besides (Marchionni, 2009). Not only did this increase users' interest and engagement in the digital resource in the UK, but also brought the initiative to a wider audience. German people were subsequently invited to contribute under the auspices of the German National Library (Jisc, 2010; Cullingford, 2011, p. 67) and the content became part of *Europeana* (*1914–18*).

Crowdsourcing could be a very positive and powerful factor in the future selection of content for digitisation. Indeed, it could be argued that it is a more immediate and direct means of meeting users' interests and needs than the distilled choices achieved through other methods. This, however, does raise issues about the intellectual authority and quality of the content. It also opens major debates about significant and sensitive societal issues for all heritage sectors (Flinn, 2010). Such matters are currently generating discussion, and it is likely to be some years before the most effective role of crowdsourcing for digitisation selection becomes clear and fully established.

4.4 THEMATIC AND SUBJECT CONTENT

If decisions about the value of digitised materials should be steered by user needs, one of the most widespread and closely related criteria likely to result is the thematic or subject coverage of the items chosen. Inevitably, this is also amongst the most diverse of criteria, given the many permutations of user constituencies; the many themes and subjects of interest to them; and the many levels of granularity, interdisciplinarity and specialism of content that might be required. There is a risk that these circumstances can complicate, if not actually obscure, a coherent picture of the disciplines and subject areas supported by digitisation. At first sight, the content presented in Chapter 3 and Appendix 1 might seem to corroborate this view. By using the broad mapping outlined earlier in this chapter, however, it is possible to identify concentrations in particular disciplines and subjects, and examples of established or potential critical mass.

At the level of broad academic disciplines, the content in Appendix 1 shows similar patterns to those identified in the Jisc survey of UK digitised materials for research 10 years ago (Bültmann et al., 2005, p. 37). The exception is content in Medicine and Biosciences, which has since been substantially increased by the Wellcome Trust's more recent large-scale initiatives (see Chapter 3). If the content in Appendix 1 is considered at this broad level, the largest concentration of material supports Arts and Humanities, followed by Social Sciences, with Medicine, Science and Technology registering far fewer initiatives and materials digitised in these fields.

At the more specific level of broad thematic or subject coverage, the subjects that appear to predominate in Appendix 1 are History, covering at least 67% of the content, Literature and Literary Studies (c12%), Visual Arts (c12%), Society and Culture (c11%), Theology and Religion (c10%), Scotland (c9%), Languages (c7%), Wales (c6%), Geography and Travel (c5%) and Music (c4%). Where more specific periods or subfields can be identified within the History category, Family History is supported by at least 16% of the projects, with c9% relating to Modern History, c6% to Local History, and c3% to Medieval History. Military, political, industrial and biographical history account for a further 13% approximately.

These estimates are based purely on the individual project entries listed in Appendix 1, whether large- or small-scale. It is significant that, although the above percentages account for many separate initiatives, critical mass does begin to emerge in certain areas. This is most immediately evident in the large-scale outputs, many of which have been digitised in stages over a number of projects and years. For example, under Family History and Society and Culture, the census data and related military and church parish records, principally from the 19th century onwards but also covering earlier periods, have been digitised by the national archive services and libraries. These have now cumulated to provide a major resource for the study of Family History, as well as supporting researchers, students and members of the wider public with an interest in many other aspects of social and military history. Similarly, the Wellcome Library's systematic digitisation of collections in the History of Medicine has provided major and wide-ranging resources for this field and will continue to expand under the organisation's current digitisation strategy. Other examples include the *House of Commons Parliamentary Papers* and the considerable number of projects to digitise newspapers (discussed more fully later), which are relevant to Modern History, as well as to the widest range of thematic and subject interests.

Two very prominent examples of strong critical mass emerge when identifying materials relating to Scotland and Wales. In both cases, a common national and cultural background draws together many resources that are also relevant to other selection criteria discussed in this chapter. In Scotland, as Chapters 2 and 3 have shown, the strategic commitments of organisations such as the National Library of Scotland (NLS) and the National Records of Scotland (NRS) have led to significant digitised collections covering Scottish culture as a whole. The NLS has provided, in selective form so far, a wide range of materials covering the unique and rare, history, language and literature, music, images, film and many more. Within this broader approach, it has focused on several large-scale initiatives, for example, conversion of its exceptionally fine map collections, and materials portraying the Scottish perspective on World War I. In parallel, the NRS has systematically digitised core resources including the Scottish Censuses, registers of births, marriages and deaths, church records, and registers of property (or sasines), along with many other categories of materials reflecting the social and cultural aspects of Scotland and its history. Relevant content in other initiatives, such as the Resource Enhancement Scheme, run by the Arts and Humanities Research Council, and NOF-Digi also contribute to this critical mass, and it will be further extended as a result of the NLS' more recently stated aim to provide digital access to all out-of-copyright Scottish publications.

If the critical mass on Scottish matters is strong, the equivalent in Wales is arguably stronger still. The early commitment of the National Library of Wales (NLW) to digitisation as a core strategic activity, as discussed in Chapter 2, has produced the range of materials described in Chapter 3 that has moved the NLW steadily towards its credible goal of comprehensive or near-comprehensive digitisation of its Welsh collections. Thus, extensive digital collections are available covering Treasures, Welsh language and literature (with a notable strength in poetry), Welsh journals and newspapers, maps, music, images, and the history of the Welsh both within the country and abroad, including World War I as experienced by the Welsh. As these have come together under the umbrella of the Theatre of Memory, other resources relating to Wales produced by other partners in CyMAL collaborations and initiatives beyond Wales complement the major initiatives within the country and extend coherent provision for the Welsh theme.

It is not only large-scale initiatives, however, that, in combination, are providing critical mass in UK digitisation. Less immediately obvious are the smaller scale projects which complement each other, whether by chance or design, and which therefore can be grouped together effectively. There are numerous examples to be drawn from Appendix 1. Triggered by the centenary of World War I, there has been a range of initiatives both large and small on this topic that combine to offer a multifaceted perspective on this major period of history. Significantly, these have not only contributed to a critical mass within the UK but also beyond, as some of the content is also available through the EU-wide initiative, *Europeana (1914–18)*. Within the Visual Arts there are common threads in the provision of photographic collections and of illustrated manuscripts from the medieval period. Medieval and early modern manuscripts, whether illustrated or not, are also the basis of the principal thread to be found in resources relevant to Theology and Religion. In Music, several separate projects have covered different aspects of and materials relating to early music and to the availability of broadside ballads from different parts of the UK.

Beyond such coherence, there are many more resources within the UK outputs that have the potential to develop into critical mass, if future selection and planning were to take account of their existence. For instance, the content relating to Local History, a strong element in the NOF-Digi programme as well as featuring more selectively in many others, indicates a rich vein for expansion to create a critical mass in relation to different regions in the country. In Literature and Literary Studies, almost all the identified examples of selection by genre (c26%) relate to poetry. English Literature is covered by c4% of the outputs, but, while much of this is diverse, some more specific concentrations are developing. For instance, just under 1% of the outputs relate to Shakespeare, including primary sources such as digitised versions of the folio and quarto editions as well as secondary sources from exhibitions and learning materials. While resources concentrating on languages do not appear in these exemplars to present clearly consolidated areas, there are the beginnings of a critical mass in relation to Welsh (c1% of projects), and some common strands are also emerging for the English language (c3% of projects).

4.5 FORMAT AND MEDIUM

Other common selection factors identified in the outputs in Appendix 1 are format and medium. While it might be argued that these do not in themselves define concepts of intellectual value, they have been prominent selection criteria since the earliest digitisation projects in the UK, as evidenced in the outputs of programmes such as eLib and NOF-Digi. Amongst the reasons that they have proved effective is that interest in multimedia resources has grown as technology has increased their potential, and because some of the materials delivered by these means are major, multidisciplinary resources of interest to many different audiences. As a result, these particular criteria represent a substantial percentage of the individual outputs in Appendix 1 (at least 23%), and they include some of the very significant large-scale digitisation projects that have now achieved critical mass. The digitisation of newspapers is a clear example. The British Library's systematic projects over many years have created the extensive *British Newspaper Archive*, and a number of smaller projects in programmes such as NOF-Digi have complemented the large-scale initiatives. Digitised maps have also developed into a critical mass, through the carefully phased projects at the National Library of Scotland, and related initiatives from the British Library, the National Library of Wales, The National Archives and other smaller outputs in NOF-Digi. By contrast, images (including paintings, drawings, etchings and photographs) are the focus of at least 12% of the projects in Appendix 1, but the initiatives are mainly small-scale and their coverage is so varied that no critical mass currently emerges from these particular examples.

4.6 COHERENCE

Coherence is a selection factor that frequently interrelates with several of the criteria already discussed earlier. With an estimate of at least 7% of relevant projects, it features less often than many other criteria. It is important, however, for ensuring that items or collections which relate less closely in terms of subject or theme are connected by a common element and therefore do combine to provide a resource of value in its totality. In their survey for Jisc and CURL, Bültmann et al. found that coherence ranked as the highest required criterion alongside the relevance of the selection to organisational aims and objectives (Bültmann et al., 2005). In practice, coherence can be achieved by applying a range of factors, including, for example, items from the same collection, publications of a particular organisation, or the work of a particular individual. Several of these are represented in the projects listed in Appendix 1, and some have now delivered critical mass. Perhaps the most prominent is the large-scale digitisation of UK Parliamentary Papers which has its origins in Phase 1 of the Follett plans, and which has been systematically developed over several years as part of later, larger digitisation programmes. Another example of critical mass based on selection for coherence is the Jisc-funded *Nineteenth Century Pamphlets Project*. Appendix 1 also includes several smaller scale projects that represent other types of coherence, such as the NOF-Digi 'Sense of Place' projects, the British Library's *Picturing Canada*, and the National Library of Scotland's *Publications of Scottish Clubs*.

4.7 VIRTUAL REUNIFICATION

Closely related to the concept of coherence is virtual reunification. This is one of the selection criteria that has come into being as a direct result of the digital environment and its potential was recognised very early on. It allows the integration of physically dispersed source materials that otherwise would remain separate, making it 'irrelevant whether the physical bits that make up a digital resource are located on a disk drive in the library or across the world' (Logoze and Fielding, 1998). Of the projects in Appendix 1, only c1% has been clearly identified in this category, but this includes some initiatives that have matured into pre-eminent resources, and have proved that virtual reunification is much valued by researchers, in particular in the Digital Humanities, but also by other, wider user groups who would not otherwise have access to such materials. For example, the *International Dunhuang Project* has brought together ancient Buddhist manuscripts, paintings, textiles and artefacts from some eight different sites, including the British Library, which has been a strong exponent of virtual reunification since the 1990s (Carpenter et al., 1998; Shenton, 2009; Calhoun, 2014, pp. 43–44). On the same principle, the earliest manuscript containing the complete New Testament as well as several books of the Old Testament has been reunited by the *Codex Sinaiticus* project (Shenton, 2009), while the *Genizah Project*, with input from the British Library, Cambridge University Library and John Rylands University Library Manchester, has allowed the renowned collection of ancient Hebrew, Arabic and Aramaic manuscripts to be reunited.

4.8 CLUSTERING

As Chapter 1 makes clear, creators and users of digitised materials have long been troubled by the 'silo' culture in digitisation, and the resulting obstacles to easy access, cross-searching and use of the resources within them. Clustering has developed as one of the current approaches to overcome these problems. It involves bringing together existing digitised resources to create an upgraded or new service, with suitable metadata and search methods for more effective use of the content. It also has relevance to the selection process for new digitisation, by allowing such work to marry new materials with content that has already been digitised. Like virtual reunification, this criterion is evident in only a small percentage of the projects in Appendix 1, with c3% clearly indicating clustering activities. It is, however, another powerful way of enhancing the value of the original materials, and outputs include some major, large-scale resources. *Connected Histories*, for example, brings together the contents of *Old Bailey Online* and the *Church of England Database*, and *Visualising China* combines a number of important photograph collections illustrating life in China between 1850 and 1950 (Jisc, 2011). These two examples show that clustering can also be applied successfully to many different types of content, as can be seen in other projects such as the British Library's *Meerut Conspiracy Files*, or the internationally based *Online Veterinary Anatomy Museum*.

4.9 FUNDERS

It has already been noted that the final selection of intellectual content for digitisation can be modified or overridden by technical, infrastructural, legal or financial factors affecting the planned work. In the financial context, experience over the last 20 years has shown that the role of funders must also be considered when selecting content for digitisation. The high costs of initiatives in the public sector have long required institutions to seek external funding to supplement, and in many cases to replace, financing from core budgets. Practices and models vary greatly, and can be dependent

on strategic requirements from the paymasters of the digitising organisations. As already noted in Chapter 3, the House of Commons Culture
Media and Sport Committee took the view in 2000 that digitisation at the
British Library should not be done at risk to the Library's core functions,
and the Library's initiatives since that time have been dependent on partnerships and third-party funders. By contrast, the National Library of Wales
and the National Library of Scotland were empowered to restructure their
core budgets to advance their digital strategies. Many more, variant permutations for combining core and external funding are to be found in the
histories of digitisation projects.

Funding for digitisation may come from international bodies, government agencies, independent trusts and foundations, nonprofit organisations,
public—private partnerships with commercial organisations and individuals,
all of whom will typically set priorities and parameters for the project to be
supported. These are sometimes very broad and general, enabling much
flexibility, but they can also be very specific about one or more aspects of
the work to be delivered. This can include the content to be digitised.
Many of the outputs in Appendix 1 have been directly influenced and
shaped by the requirements of the special funding programmes under
which they were produced, or by funding bodies or individuals who have
supported particular content of interest to them for educational, professional or personal reasons. In the most extreme circumstances, a funder's
priorities can necessarily take precedence over all the other selection criteria
discussed earlier, although there are also many instances where funders will
work in partnership with the digitising organisation to establish mutually
beneficial choices. In general, however, it is wise to include funders' requirements amongst the selection criteria being applied when initiating a
digitisation project, and to consider sharing with the funder the organisation's criteria for selecting and prioritising the target content, to aid mutual
understanding and effective collaboration.

4.10 DEVELOPMENT OF CRITICAL MASS

In this chapter, principal core criteria for the selection of content to be
digitised have been identified from digitisation guidelines developed over
the last 20 years, and these are reflected in the outputs of the UK projects
covered in Chapter 3 and Appendix 1. An indicative mapping of these
projects against the criteria also suggests that separate and in many cases
apparently unconnected UK outputs do interrelate in ways that offer actual

or potential critical mass at national level. This suggests that there is scope for more effective coordination at national level and this issue is explored in the following chapter.

REFERENCES

Ayris, P. (1998) *Guidance for Selecting Materials for Digitisation.* (Joint RLG and NPO Preservation Conference: Guidelines for Digital Imaging). [Online] Available from: http://eprints.ucl.ac.uk/492/1/paul_ayris3.pdf.

Birrell, D., Dobreva, M., Dunsire, G., Griffiths, J., Hartley, R. and Menzies, K. (2009) DiSCmap: digitisation of Special Collection: mapping, assessment, prioritisation: final report. [London]: JISC. [Online] Available at http://www.webarchive.org.uk/wayback/archive/20140614060147/http://www.jisc.ac.uk/whatwedo/programmes/digitisation/reports/discmap.aspx#downloads.

Bültmann, B., Hardy, R., Muir, A. and Wictor, C. (2005) *Digitised Content in the UK Research Library and Archives Sector: a report to the Consortium of University and Research Libraries; and the Joint Information Systems Committee.* [Bristol?]: JISC; CURL. [Online] Available from: http://www.jisc.ac.uk/uploaded_ documents/JISC-Digi-in-UK-FULL-v1-final.pdf.

Calhoun, K. (2014) *Exploring Digital Libraries.* London: Facet.

Carpenter, L., Shaw, S. and Prescott, A. (Eds.) (1998) *Towards the Digital Library: the British Library's Initiatives for Access Programme.* London: British Library.

Cornell University Library. (2004) *Cornell University Library Digital Preservation Policy Framework.* [Online] Available from: http://ecommons.library.cornell.edu/bitstream/1813/11230/1/cul-dp-framework.pdf.

Cullingford, A. (2011) *The Special Collections Handbook.* London: Facet.

Daigle, B. (2012) Stewardship and curation in a digital world. In Fieldhouse, M. and Marshall, A. (Eds.) *Collection Development in the Digital Age.* London: Facet: 93–107.

Digital Library of Georgia. (2001) *Digital Library of Georgia Digitization Guide.* [Online] Available from: http://dlg.galileo.usg.edu/guide.html.

DigitalNZ. (2009) *Selecting for Digitisation.* [Online] Available from: http://www.digitalnz.org/make-it-digital/selecting-for-digitisation.

Dobreva, M., O'Dwyer, A. and Konstantelos, L. (2012) User need in digitization. In Hughes, L.M. (Ed.) *Evaluating and Measuring the Value, Use and Impact of Digital Collections.* London: Facet: 73–84.

Erway, R. (2008) Supply and demand: Special Collections and digitisation. *LIBER Quarterly.* Vol. 18, No. 3/4: 324–36.

ESYS Consulting. (2000) *Summative Evaluation of Phases 1 and 2 of the eLib Initiative: Final Report.* Guildford: ESYS. [Online] Available from: http://opus.bath.ac.uk/35005/1/elib_fr_v1_2.pdf.

Federal Agencies Digitization Guidelines Initiative. (2009) *Digitization Activities: project planning and management outline.* [Online] Available from: http://www.digitizationguidelines.gov/guidelines/digitize-planning.html.

Flinn, A. (2010) An attack on professionalism and scholarship?: democratising archives and the production of knowledge. *Ariadne.* Issue 62. [Online] Available from: http://www.ariadne.ac.uk/issue62/flinn.

Gertz, J. (2013) Should You? May You? Can You? *Computers in Libraries.* Vol. 33, No. 2: 7–11.

Hazan, D., Horrell, J. and Merrill-Oldham, J. (1998) *Selecting Research Collections for Digitization: Full Report.* Council on Library and Information Resources. (Publication 74). [Online] Available from: http://www.clir.org/pubs/reports/hazen/pub74.html.

Hughes, L. M. (2012) Introduction: the value, use and impact of digital collections. In Hughes, L. M. (Ed.) *Evaluating and Measuring the Value, Use and Impact of Digital Collections.* London: Facet,1-10.

IFLA. (2002) *Guidelines for Digitization Projects for Collections and Holdings in the Public Domain, particularly those held by Libraries and Archives.* [Online] Available from: http://www.ifla. org/files/assets/preservation-and-conservation/publications/digitization-projects-guidelines.pdf.

Jisc. (2010) *Great War Archive rolls out in Germany.* [Online] Available from: https://www. jisc.ac.uk/news/great-war-archive-rolled-out-in-germany-16-dec-2010.

Jisc. (2011) *Clustering and Sustaining Digital Resources: the JISC eContent Programme 2009-2011.* [Online] Available from: http://www.webarchive.org.uk/wayback/archive/20140615013612/http://www.jisc.ac.uk/media/documents/publications/general/2011/JISCeContentClusteringAndSustainingDigitalResources.pdf.

Jisc Digital Media. (2015) *High Level Digitisation for Audiovisual Resources.* [Online] Available from: http://jiscdigitalmedia.ac.uk/infokit/audiovisual-digitisation/audiovisual-digitisation-home.

Johnson, P. (1997) Collection Development Policies and Electronic Information Resources. In Gorman, G.E and Miller, R.H. (Eds.) (1997) *Collection Management for the 21st Century: a handbook for librarians.* Westport: London, Greenwood: 83—104.

Lee, S. D. (1999) *Scoping the Future of the University of Oxford's Digital Library Collections: final report.* [Online] Available from: http://www.bodley.ox.ac.uk/scoping/report.html.

Lesk, M. (2005) *Understanding Digital Libraries.* San Francisco: Morgan Kaufman.

Library of Congress. (n.d.) *Preservation Digital Reformatting Program: Selection Criteria.* [Online] Available from: http://www.loc.gov/preservation/about/prd/presdig/presselection.html.

Logoze, C. and Fielding, D. (1998) Defining collections in distributed digital libraries. *D-Lib Magazine,* Vol. 4, Issue 11 [Online] Available from: http://www.dlib.org/dlib/november98/lagoze/11lagoze.html.

Marchionni, P. (2009) Why are users so useful?: user engagement and the experience of the JISC Digitisation Programme. *Ariadne.* Issue 61. [Online] Available from: http://www. ariadne.ac.uk/issue61/marchionni.

Maron, N.L. and Pickle, S. (2013) *Appraising our Digital Investment: sustainability of digitized Special Collections in ARL libraries: a report from Ithaka S+R and the Association of Research Libraries.* [Online] Available from: http://sr.ithaka.org/sites/default/files/reports/digitizing-special-collections-report-21feb13.pdf.

McMenemy, D. (2005) Creating digitized content in community libraries. In McMenemy, D. and Poulter, A. *Delivering Digital Services: a handbook for public libraries and learning centres.* London: Facet.

National Library of Australia. (c2012) *Collection Digitisation Policy.* [Online] Available from: https://www.nla.gov.au/policy-and-planning/collection-digitisation-policy.

National Library of Wales. (2009) *Digitisation Strategy 2008/9 — 2010/11.* [Online] Available from: http://www.llgc.org.uk/fileadmin/fileadmin/docs_gwefan/amdanom_ni/dogfennaeth_gorfforaethol/dog_gorff_strat_dig_08_09_10_11S.pdf.

National Library of Wales. (2011) *Digitisation Strategy 2011/12—14/15.* [Online] Available from: http://www.llgc.org.uk/fileadmin/fileadmin/docs_gwefan/amdanom_ni/dogfennaeth_gorfforaethol/StrategaethDdigido2012-2015.pdf.

NEDCC. (c2007) *NEDCC Preservation Leaflets: Reformatting 6.6 Preservation and Selection for Digitization.* (Author: Janet Gertz). [Online] Available from: https://www.nedcc.org/free-resources/preservation-leaflets/6.-reformatting/6.6-preservation-and-selection-for-digitization.

NINCH. (2003) *The NINCH Guide to Good Practice in the Digital Representation and Management of Cultural Heritage Materials.* [Online] Available from: http://www.ninch.org/programs/practice.

NISO. (2007) *A Framework of Guidance for Building Good Digital Collections* (3rd ed.). [Online] Available from: http://www.niso.org/publications/rp/framework3.pdf.

Ooghe, B. and Moreels, D. (2009) Analysing Selection for Digitisation. *D-Lib Magazine.* Vol. 15, Issue 9/10. [Online] Available from: http://www.dlib.org/dlib/september09/ooghe/09ooghe.html.

Rusbridge, C. (2001) After eLib. *Ariadne.* Issue 26. [Online] Available from: http://www.ariadne.ac.uk/issue26/chris/.

Shenton, H. (2009) Virtual reunification, virtual preservation and enhanced conservation. *Alexandria.* Vol. 21, No. 2: 33—45.

Shuttleworth, S. (2015) Old Weather: citizen scientists in the 19th and 20th centuries. *Science Museum Group Journal.* Spring 2015. [Online] Available from: http://journal.sciencemuseum.ac.uk/browse/issue-03/old-weather/.

Strategic Content Alliance. (2012) *Audience Development and Engagement Research.* [Online] Available from: http://sca.jiscinvolve.org/wp/allpublications/audience-publications/.

Tanner, S. (2011) The value and impact of digitized resource for learning, teaching, research and enjoyment. In Hughes, L. M. (Ed.) *Evaluating and Measuring the Value, Use and Impact of Digital Collections.* London: Facet: 103—120.

UNESCO. (n.d.) *Fundamental Principles of Digitization of Documentary Heritage.* [Online] Available from: http://www.unesco.org/new/fileadmin/MULTIMEDIA/HQ/CI/CI/pdf/mow/digitization_guidelines_for_web.pdf.

University of California Libraries. (2013) University of California Selection Criteria for Digitization (PAG). [Online] Available from: http://libraries.universityofcalifornia.edu/content/university-california-selection-criteria-digitization-pag.

Warwick, C. (2012) Studying users in digital humanities. In Warwick, C., Terras, M. and Nyhan, J. (Eds.) *Digital Humanities in Practice.* London: Facet: 1—21.

Warwick, C., Terras, M. and Nyhan, J. (2012) Introduction. In Warwick, C., Terras, M. and Nyhan, J. (Eds.) *Digital Humanities in Practice.* London: Facet: xiii-xix.

Yale University Library. (2006) *Report of the Working Group for Developing Selection Criteria for Collections Digitization.* [Online] Available from: http://www.library.yale.edu/iac/documents/digcriteria.final.html.

CHAPTER 5

The Future for Collaboration

Whatever advances might presently be made in the UK's strategic collaboration to support publicly and philanthropically funded digitisation, progress will depend on the past and current circumstances described in earlier chapters. Digitisation since the 1990s, in the UK as in some other countries, has developed in several key stages which might reasonably be looked on as 'peaks and troughs'. An energetic and visionary start initiated ambitious and wide-ranging work to transfer analogue content to digital form and to establish the new practices and standards required by the digital environment. This was followed by a period of consolidation and expansion, before financial pressures of various sorts led to reduced funding opportunities for many of the key players. In recent years, the consequent limitations on the scale and range of further work have been very marked, and at the time of writing there seems little prospect of this being reversed in the foreseeable future. These trends are fully reflected in the purpose and outputs of the exemplar projects outlined in Chapter 3 and Appendix 1, as is the poorly coordinated nature of UK outputs. At first sight, Chapter 3 and Appendix 1 seem to confirm the views cited in Chapter 1 that, in common with digitisation work in many other countries, overall provision in the UK has been created in silos, and is disparate and piecemeal.

This situation seems likely to continue, and is exacerbated by the complex strategic environment within which digitisation efforts take place. As Chapters 2 and 3 demonstrate, digitisation strategies have varied in their strategic prominence and detail. Whatever their status, they have all been subsets of the wider strategies of many organisations and agencies. To complicate the picture further, these strategies have operated at different, but interrelated, levels. Both UK and devolved governmental digital strategies have steered the digitisation strategies of local government, and of educational and publicly funded heritage agencies, who have in turn influenced the strategic choices of individual organisations within their sector. Alongside this hierarchy, the philanthropic sector has pursued priorities and initiatives that support their independent perspectives, but that often have much in common with the interests of the publicly funded organisations.

Stepping Away from the Silos
ISBN 978-0-08-100278-0
http://dx.doi.org/10.1016/B978-0-08-100278-0.00005-2

The most positive element in this scenario and the continuing strength for the UK is the long-term commitment shown by educational, heritage and learned organisations, despite changed priorities in other organisations' strategies. Their forward planning continues to place digitisation in a central and essential role. They have become the keepers of the future of digitisation in the public and philanthropic sectors, as wider political and societal interests have shifted to the promise of newer and emerging technologies. It is also to the good that collaborative approaches have been universally supported at all strategic levels. This has generated and optimised major financial resources, supported essential infrastructural development and facilitated a moderate level of coordinated, cross-sectoral work, but it has by no means overcome the disparate nature of initiatives, and of the resulting resources. New initiatives have repeatedly aimed to contribute to a wider picture, but it is a picture that has frequently been ill-defined, and in some instances completely obscure. This scenario suggests the need for a national digitisation strategy to set the context and priorities for further digitisation work.

5.1 ENVISIONING A UK NATIONAL DIGITISATION STRATEGY

In raising this question, it is important to consider whether the role of digitisation remains sufficiently important to merit such a major and challenging venture. The field has lost its prominent position in strategic planning because the digital world has moved on, in that inexorable way that has consigned many exciting new information technology developments to the ranks of routine processes after a few years. Born-digital is now a normal means of producing resources in all fields of activity, whether or not they are subsequently made available in a physical format. Digital resources are increasingly complex and creative; they are not just consulted on screen, but are used, shared, reused and reshaped to create more new resources. It is not unreasonable to argue that the isolated activity of digitising analogue content is no longer enough. In keeping with this, many of the more recent projects that have supported digitisation have laid greater emphasis on deliverables such as wider intellectual content emerging from the new resource, and other types of added value, such as potential reuse of the content or specialist services that depend on the underlying content.

Nonetheless, this new era of creation and creativity sits over and depends on the fundamental activity of converting analogue content to digital. As Green observed 'digitisation is an ugly word for a beautiful concept…the digital has become the kingdom in which many people live much of their intellectual and even social lives' (Green, 2009, p. 1). Potentially there are undesirable consequences of not continuing to address digitisation. As noted in Chapter 1, the same or similar content may be created anew in born-digital form, an unnecessary duplication of effort. More seriously, there is the risk that as users become ever more digitally based, knowledge and information in existing analogue formats will be overlooked or actually lost.

Given this context, it is worth considering how far the UK might improve its national coordination, and whether a national strategy could be developed. Such a proposal, however, would be far from new in the UK. There have already been several calls for such a development since the late 1990s at least. One of the earliest came in a major study carried out by the Humanities Advanced Technology and Information Institute (HATII) in 1997 to inform the Heritage Lottery Fund's (HLF) planning for their distribution of the New Opportunities Fund (Ross & Economou, 1998; Ross et al., 1998). Alongside a range of recommendations for the HLF's sole attention, the report stressed the need for a national strategy for ICT in the heritage sector, to be referred to the Department for Culture, Media & Sport for further action. The strategy's purpose would be to maximise the development and benefits of digital heritage resources and to support organisations who were calling for such an initiative at the time. These included archives, biological recording centres, countryside organisations, libraries and museums. The focus was both on digitisation of analogue materials, and on born-digital outputs which were growing in number and importance at that time. The proposed coverage of such a strategy was wide-ranging: infrastructural issues, interoperability, software development, resourcing, preservation, management of the outputs, commercial exploitation, and enabling access to all organisations and sections of society for their benefit in their education and their employment. A key development identified for this strategy was a national digital collection that would develop critical mass over time, based on a national heritage network supporting synergy between projects, whether for digitised or born-digital content.

While the HATII report addressed a digital strategy for the heritage sector as a whole, the higher education sector also turned to the matter, focusing more closely, however, on libraries and research provision. The Research Support Libraries Group (RSLG), instigated in 2001 under the chairmanship of Sir Brian Follett, concentrated specifically on developing a national strategy to enable ready and unimpeded access to research information in all its forms for the UK's research community. The Group's terms of reference included seeking a national strategy for digitising existing collections of primary research material (Follett, 2001). The Group's deliberations were lengthy, and their consultations were very wide-ranging. While their final report proposed major developments for research support at a national level that were duly implemented, it was clear that the priorities had shifted as a result of the Group's consultations, and there was no explicit proposal for a national digitisation strategy.

In 2002, during the period of the RSLG's work, Pearson considered the current progress in digital content creation, and digitisation in particular, by the library sector, which he saw as being 'in such a rudderless state...; no shortage of action, but no overall sense of direction' (Pearson, 2002). Here again there is a strong emphasis on the need for an agreed national strategy, to provide a context for decision making. Pearson suggests the fundamental principles for such a strategy, in this case with the focus on the content to be delivered rather than on technical and infrastructural matters. This could be based on agreement about the proportion of the documentary heritage requiring digitisation within an agreed, extended timeframe, and prioritisation of content according to whether entire collections or selections of them were to be converted. Once digitised, the content could be accessible on a distributed basis, with a central database carrying the metadata for the relevant sites.

By 2004, the British Library was well established as a leader in digitisation, and this was recognised by the National Audit Office (NAO) in its report on the Library's services (National Audit Office, 2004). The report noted, nevertheless, the absence of national oversight for digitisation, and the attendant risks of incompatible technical standards, of duplication, and of failure to learn the lessons from digitisation activities. It did not go so far as to recommend a national strategy to address this, however, calling instead for collaboration between the various government bodies involved in digitisation to be a requirement of their formal agreements for government funding.

The NAO's observations were part of the wide range of evidence informing Jisc's review of digitisation in the UK, published in the following year (Jisc, 2005). This cross-sectoral study examined in some detail the evidence of UK practice in all the principal aspects of digitisation up to 2005, from guidance on standards to comprehensive listings of projects undertaken. Its overall findings were in keeping with previous reports and commentaries. There was a lack of coordination amongst all constituent parts of the digitisation process in the UK, and the evidence showed a great need to address this problem. This study, however, did not propose establishing a strategy *per se*, but rather the development of a national framework for digitisation, as part of a UK e-Content Strategy, that would be flexible, issuing clear guidelines but not prescriptive requirements, and applicable in a distributed rather than centralised environment. This 'Digitisation Framework' was expected to help to fill gaps in provision, to reduce overlaps and to support the provision, take-up and use of digitised resources.

Although this proposed Framework did not emerge as a separate entity following this report, some elements of its recommendations were to be found in the work of the Strategic Content Alliance (SCA) (Jisc, 2012), which ran under Jisc's leadership from 2006 to 2013, after which the Wellcome Library took over the lead for specific aspects of the organisation's later investigations. This collaboration has perhaps been the closest that the UK public and not-for-profit sectors have come, so far, in attempting to establish a coordinated, cross-sectoral approach to digital content provision, covering both born-digital material and digitisation from analogue sources. Between 2006 and 2013, it aimed to maximise the financial and intellectual investment in digital content collections, and to overcome the acknowledged fragmentation by developing a framework that would improve coordination between existing outputs and new initiatives. It brought together major and very influential participants: Jisc, the Arts Council England, the British Library, the BBC, the Heritage Lottery Fund and the Wellcome Library. All were deeply involved in the creation, management and exploitation of digital content for the common good. The SCA's Content Framework, issued in 2009, provided detailed guidance in audience analysis and impact; business modelling and sustainability; intellectual property rights and licensing; and maximisation of online resources (Jisc, 2011). These became the main body of work for the community, with the focus shifting to their implementation from 2009 onwards. The SCA's more recent work, now under the Wellcome's leadership, has been a wide and detailed consultation on the value of and potential for creating a Digital Public Space for the UK (Sero HE, 2014). This complex and fluid concept focuses on developing an infrastructure, in its widest sense, that would open up a critical mass of the UK's publicly funded digital content for wider purposes and audiences than is currently the case. This would, of course, include the type of digitised content under consideration in this book. The final report makes clear that, although the scale and complexities of such a venture preclude an early commitment to its development, the momentum of the SCA's interest and experience should be exploited by commissioning a 'Proof of Concept' project. This would strengthen strategic collaboration in the UK by bringing together the SCA members with other key UK players in the provision of digital content. Should this have a positive outcome, however, there will still be many years of development before this ambitious vision would become a reality.

During the same period that the SCA was creating its Framework, *Digital Britain*, the UK Government's new plans for the country's digital future as outlined in Chapter 2, was in preparation. The evidence for the report included an extensive consultation with leaders in all relevant sectors, both public and commercial. The British Library entered into the discussions and recognised that this was an important opportunity to raise awareness at government level of the need for a coherent approach to digital content creation and related aspects. Dame Lynne Brindley, as Chief Executive of the Library, called for a coherent UK national digital strategy. It was made clear that this needed to include mass digitisation of content, as well as ensuring that the wider population developed the digital literacy skills required to benefit from the extending digital environment that the plans extolled (British Library, 1999–2015, see 2008–09, p. 17).

While the SCA has been exploring its long-term vision for digital content in its widest sense, Research Libraries UK (RLUK) has revisited the role and potential for coordinated digitisation by concentrating explicitly on the future for large-scale digitisation in higher education and particularly in academic libraries supporting research (Pressler, 2014). Like the fuller studies of earlier years, this report examines many facets of the digitisation process, whether technical, legal, operational, economic, timetabling, financial and cultural. It concludes that the UK now has the relevant technical expertise to support the development of a national digitisation strategy, but may not yet be fully secure in the areas of preservation and storage. Specifically, it suggests that RLUK develop a national digitisation strategy for all its members' historic collections, an initiative which could start, if necessary, by selective digitisation, to be developed into a comprehensive coverage over time.

5.2 NATIONAL DIGITISATION STRATEGIES BEYOND THE UK

5.2.1 European Union

Far from being unique to the UK, recognition of need for a national strategy for digitisation is an international issue. It has followed many different courses and debates in different parts of the world. The European Union (EU) has an overarching strategy that embraces the concept, and several countries in and beyond the EU are currently taking active steps to progress strategies, in various ways.

Europe 2020, the European Union (EU)'s current strategy, sets objectives for EU growth to 2020. These are based on seven pillars, one of which is the Digital Agenda for Europe (European Commission, n.d.). This strand's purpose is to foster innovation, economic growth and progress by better exploiting the potential of ICT. Its principal areas of interest are the Digital Single Market; interoperability and standards; trust and security; fast and ultrafast Internet access; research and innovation; enhanced digital literacy, skills and inclusion; and ICT-enabled benefits for EU society. Significantly, this last element includes explicitly the digitisation of content through *Europeana*. This goal is no chance appearance, but the latest manifestation of a commitment that goes back at least to the Lund Principles of 2001 (European Commission, 2001). These included the development of a European view on digitisation policies and programmes, and collaboration to make digitised cultural and scientific heritage of Europe visible and accessible. Amongst a wide range of subsequent European work in this field, the MINERVA programme followed (MINERVA, 2013) along with the emergence of the *European Digital Library* and *Europeana*. The current position has developed from *New Renaissance*, the report of the Comité des Sages issued in January 2011 (European Commission; Comité des Sages, 2011). This gave new impetus to the EU's support for digitisation of Europe's cultural heritage. It set priorities at European level for digitisation and access to content, not by selection but on the principle of quantitative targets, the outputs to be accessible through the single access point of *Europeana*. A supporting set of recommendations (European Commission, 2011) introduced a target of 30 million objects to be digitised by 2015 as 'a stepping stone for getting Europe's entire cultural heritage digitised by 2025' (European Commission, 2011; para 15).

The EU, therefore, communicates an ambitious goal for European digitisation, and has set an overarching strategic context for its achievement, but it is the responsibility of Member States to devise their nation's strategy and to implement it for delivery to the people of the European Community. The Member States' Expert Group (European Commission, 2015) monitors developments. Its most recent report shows 'certain progress' but 'digitisation remains a challenge', with acknowledgement that only a small fraction of collections has been digitised so far, and additional effort is needed (European Commission; Directorate-General for Communications Networks, Content and Technology, 2014, p. 6). *Europeana* has reached its target of 30 million digitised objects available online ahead of time, but many issues remain concerning the non-digitised materials (European Commission; Directorate-General for Communications Networks, Content and Technology, 2014, p. 7). As far as national strategies go, these exist, but the report identifies such a variety of models that no coherent approach can be identified (European Commission; Directorate-General for Communications Networks, Content and Technology, 2014, pp. 8—13). Across the EU, the different countries' strategies are also at many different stages of maturity, and a recent report to the French government suggests that some have chosen to pause their activity for various political or financial reasons (France; secrétariat général pour la modernisation de l'action publique, 2014). Significantly, they are mostly developed and managed at institutional, regional or sectoral level (European Commission; Directorate-General for Communications Networks, Content and Technology, 2014, pp. 8—13; France; secrétariat général pour la modernisation de l'action publique, 2014, p. 3), in over two-thirds of cases without a formally stated digitisation policy (France; secrétariat général pour la modernisation de l'action publique, 2014, p. 11). Overall, the coverage of digitisation policies and/or programmes in the EU nations varies considerably depending on the circumstances and issues in each country.

5.2.2 Sweden

Amongst the closest to a clearly defined and coherent national digitisation strategy is the initiative in Sweden, where the government has taken a proactive approach to motivate the cultural institutions under its control. In 2011, their ICT strategy to support their response to the Digital Agenda for Europe was approved (Government Offices of Sweden, 2011). This addressed the wide range of issues typical of many government strategies in the last ten years, including *Digital Britain* and its successor documents, as outlined in Chapter 2. Thus, it gives directions for developments in infrastructure, broadband provision, the digital economy, digital inclusion, services and the societal benefits of ICT. It also goes beyond the passive interest in digitisation expressed in the UK government's strategy. It requires all central government institutions that collect and preserve cultural heritage information and make it accessible to draw up plans for digitisation and accessibility. It also calls for greater coordination of the work of relevant institutions. Some months ahead of this strategy's approval, the government had already tasked the National Archives of Sweden with setting up a secretariat to coordinate digitisation, digital preservation and digital accessibility. Now functioning as Digisam, it is led by the National Archives, the National Library of Sweden, the Council of Swedish Central Museums and the Swedish National Heritage Board. Each institution still carries out its own digitisation, but a recent evaluation of the Digisam initiative has reported positive progress overall, with improved coordination and collaboration in all aspects of the work, including prioritisation of work to be done (Swedish Agency for Public Management, 2014). It also advises firmly that coordination needs to continue. Digisam is scheduled to run until 2015, when its team must make proposals for the next stages of digitisation in Sweden. The evaluation has recommended that Digisam be tasked to maintain work so far developed while the government decides its position on the proposals for the future.

5.2.3 France

Since the appearance of the EU's progress report (European Commission; Directorate-General for Communications Networks, Content and Technology, 2014), the French Ministry of Culture and Communication has carried out the review announced in the report. Up to 2012, France had already been very active in the digitisation of the country's cultural heritage, with the government supporting a series of digitisation plans (*plans de numérisation*) for different heritage sectors (Dalbera & Pascon, 1999). Qualitative and/or quantitative targets were specified for these in multi-annual contracts between the Ministry and the institutions, which included the Institut national de l'audiovisuel (Ina), the Bibliothèque nationale de France (BnF) and its high profile development of *Gallica*[1] and the Centre national du cinéma et de l'image animée (CNC) (European Commission : Directorate-General for Communications Networks, Content and Technology, 2014, p. 12). As a result of the detailed investigations of the recent review into the digitisation of cultural materials in France and its recommendations (France; Ministère de la Culture et de la Communication, 2015a), the Ministry is aiming to set a national strategy for digital development and practice, building on the activities of the past 10 years and focusing on digitisation of books, the press, cinema and audiovisual materials in France, to be taken forward in the context of the new digital environment and practices now developing (France; Ministère de la

---

[1] http://gallica.bnf.fr/.

Culture et de la Communication, 2015b). It is anticipated that the policy will set a framework and guidelines that allow participating institutions to make informed decisions about matters such as what to digitise, for whom and why; technical and infrastructural standards to be used; metadata; copyright management; shared systems; preservation practices; promotion of content; financing of initiatives; required skills and evaluation practice. In the first *Action Plan*, issued in 2015 (France; Ministère de la Culture et de la Communication, 2015c), there are two strategic principles: a shift in digitisation by participating institutions to better integrate practices and services, and attention to copyright issues to enable accessibility and reuse of cultural content. The Département des programmes numériques (DPN) within the Ministry's secrétariat général is to have formal and active oversight of the initiative, and the participants will be the BnF, Ina, the Agence photo de la Réunion des musées nationaux (RMN), archives and the CNC. The strategy is expected to enable identification at Ministry level of collections to be included in a prioritised digitisation programme from 2016 onwards; the financial resources required; decisions about enhance-ment of existing and future digitised resources; the development of services, especially based on semantic web technologies; and the reuse and pro-motion of digitised cultural resources linked to the ecosystem of creators and innovators. The strategy will be further developed and amplified by a ministerial committee.

5.2.4 Norway

The EU is not the only part of the world where there have been efforts to set national strategies for digitisation. In recent years, notable developments include those in Norway and in Australia.

The developments in Sweden have many similarities to the very strong position taken in Norway just a few years earlier. In 2009, the Norwegian Ministry of Culture issued their National Strategy for Digital Preservation and Dissemination of Cultural Heritage (Norwegian Ministry of Culture, 2010). This expanded on the Norwegian Government's ICT policy (Norwegian Ministry of Government Administration and Reform, c2007) in relation to the cultural sector. It sets the vision 'to make as much as possible of the collections in our archives, libraries and museums accessible to as many as possible through future-oriented technological solutions' (Norwegian Ministry of Culture, 2010, p. 7). Significantly, it gave priority to digitisation to achieve this, and recognised Norway's very real opportunity for this to be advanced by building on the number of major digitisation initiatives in the area of cultural heritage that had already been developed in the country.

The Ministry stated clearly that it wanted to put in place shared access to content and services in Norwegian libraries, archives and museums. The strategy therefore proposed as its key aim to make a targeted central government effort to create the basic conditions that would 'foster easy and adequate digital access to cultural material in all its diversity' (Norwegian Ministry of Culture, 2010, p. 9). As well as addressing the many established elements of a digitisation strategy, including common standards, IT systems, storage, preservation, dissemination, search services across the sectors and within each material type and expertise, it also supported the development of programmes for mass digitisation. By this time, the National Library had already embarked on its long-term programme to digitise all its Norwegian material (Takle, 2009), and the National Archival Services of Norway had created a Digital Archive of selected registers, rolls and official registration material, as well as folk narratives, reflecting the history of the Norwegian population. However, while the museum sector had made progress in the universities, the wider sector was seen as needing to engage more fully in digitisation.

To progress its vision, the Ministry required the National Library of Norway, the National Archival Services of Norway and the Norwegian Archive, Library and Museum Authority to take responsibility for setting and implementing a digitisation strategy for libraries, archives and museums, respectively. The Ministry also made very clear that all participants must pull in the same direction: 'collaboration and interaction are to be the rule, not the exception' (Norwegian Ministry of Culture, 2010, p. 13). Formal oversight for this would be carried out by a newly formed Norwegian Digitisation Council.

This approach has many similarities to the developments already described in Sweden. In Norway's case, the Government's vision and commitment is particularly strongly expressed. Beyond the evident recognition of the benefits of nationally structured digitisation activity, this may also owe its momentum to the Norwegian cultural tradition of '*almenning*', or 'owned by all' (Takle, 2009), a concept that would undoubtedly help governments, institutions and the general population to understand the role of digitisation in enabling democratisation of heritage resources.

5.2.5 Australia

In Australia, there have been some major influential digital initiatives over some years arising from national collaborative efforts, including for example Trove,[2] the National Library of Australia's aggregator and repository for digitised materials on Australia. Recently, leading figures from institutions and organisations at all levels of the GLAM sector (galleries, libraries, archives and museums) have joined together to examine the opportunities and challenges created by new broadband and digital services, as they relate to the GLAM sector, and how these organisations should move forward. The result is a carefully structured, detailed and highly consultative *Innovation Study* (Mansfield et al., 2014) with constructive outcomes. In the participants' view, Australia has displayed disparate responses and initiatives in response to the digital environment, that have been manifested in a 'current mix of resistance, ignorance, piecemeal adoption and in some cases wholesale embracing of digital' (Mansfield et al., 2014, p. v). A programme of consultations, workshops and post-workshop interviews produced proposals for four strategic initiatives for cross-sectoral development: making the public part of what we do; becoming central to community wellbeing; beyond digitisation — creative reuse; and developing funding for strategic initiatives. This approach to national coordination is expressed in terms that carry notably different emphases and priorities to equivalent strategies in other countries, such as those mentioned earlier, and it therefore offers a fresh perspective on many underlying issues that are also at the heart of other countries' deliberations.

The suggested strategic initiatives are supported in the study by an action plan of specific developments to be progressed within ten, five or two years (Mansfield et al., 2014, pp. 70—74). These are proposed to be taken forward in the context of a National Framework for Collaboration, which is defined as covering digitisation and access; digital preservation; national approaches to rights; skills and organisational change; shared infrastructure; and transdisciplinary collaboration and research partnerships. It is made clear, however, that the four strategic approaches will require coordination and a voice to take them to government, as there is currently no formal gathering for leaders or practitioners across the GLAM sector. The study proposes a National Leadership and Collaboration Forum to address this matter.

---

[2] http://trove.nla.gov.au/.

These developments in Australia present an interesting variant on the models for national collaboration identified previously in some European countries. Essentially, a similar grouping of institutions across the heritage sector are addressing digital issues, with a strong focus on digitisation, but in this case the initiative and momentum come from the GLAM professionals rather than government. Moreover it embraces participants from large and small institutions and goes beyond the institutions under the direct steer of government.

5.3 REALISING A UK NATIONAL DIGITISATION STRATEGY

The UK digitisation communities have long understood the potential of a national digitisation strategy and have grappled with the challenges of making it a reality. In theory, an overarching strategy would engender levels of coordination that would in turn produce a coherent range of digitised resources. This would present the existing content clearly to its users from all communities, would highlight gaps that need to be filled, and would enable more targeted planning and funding of further digitisation. In practice, a full national strategy is a multifaceted exercise, embracing several major areas of strategy, policy and operations. As Chapters 2 and 3 show, many different UK players need to be fully aligned with and committed to the aims before it can be initiated. Such circumstances are equally true of many other countries. The recent developments in Europe and Australia as discussed earlier would indicate, however, that some clarity is emerging and that in these nations at least there is a momentum to make progress towards planned and coordinated digitisation of analogue materials.

There are important common factors in these examples. In Sweden, Norway and France, there is a clear political will at government level to take such an initiative forward, with the attendant authority to require delivery by key players and to evaluate their results. Smaller nations appear to have been able to take action faster and more confidently. It is noticeable that the momentum has developed earlier and more strongly in Norway and Sweden. This is probably due in part to the more manageable scale of the venture, with fewer stakeholders involved and users to be served. It is therefore interesting to see that a much larger nation like France is now choosing a similar approach, and their experience and developments are likely to offer insights for other larger nations.

If the size of the country is different, however, there is the common factor that all these countries have their own distinct cultural and language base, and, while not excluding activity in other areas, the national strategies are expected to prioritise these aspects for digitisation. A further common factor is that the necessary planning and action are to be taken by organisations that are all leaders in their particular heritage sectors but that also answer directly to government and are consequently obliged to respond to the overall strategy. To this extent, therefore, the nascent national strategies do not necessarily draw in the wider heritage, educational and learned organisations in their countries. By contrast, the Australian initiative does indeed embrace these wider communities, but without the active commitment of government to support planning and progress.

If the national context of the aforementioned examples is not identical to the UK's, there are nonetheless some parallels. As discussed in Chapter 2, the overall political will is weak and at the level of the UK government is especially reactive. The Welsh and Scottish governments have taken a more engaged approach. Like those of Sweden, France and Norway, they have focused on the cultural importance of digitising their nations' heritage, and have recognised this within their overall strategic planning, but they still fall short of supporting formal national strategies. While these devolved positions might be helpful building blocks for wider national coordination in the UK, they are not sufficient, currently, to unite the many relevant stakeholders, such as those covered in Chapter 2. All those organisations are operating in too complex a matrix of strategy-making in the UK and without sufficient political commitment at government level to overcome the many variant strategic interests at play. If further coordination is sought in the UK, it is most likely to fall to the heritage and educational sectors to initiate it and seek the necessary government support, as has been identified in Australia in the current time.

5.4 CAPITALISING ON COLLABORATIVE CULTURE

It is therefore important that these players recognise their central role in driving strategy for future digitisation, whatever form it might take. As already noted, in the UK they start with the advantage of an existing culture of collaboration. There could be value in considering how this might be further strengthened to identify what national coordination is required and is practical. In the current climate, the scope and scale of a full-scale digitisation strategy in the multilevel, multifaceted strategic context of the UK does seem too challenging to tackle, and the absence of progress in past years, despite authoritative calls for action, would seem to endorse this.

There are too many players with too many varying strategic interests. Any resulting strategy is likely either to be too vague, in an effort to engage all participants, or too restrictive, thereby stifling creativity and innovation or simply failing to engage the relevant organisations.

If there is a will to achieve better coordination, it seems more likely that this can be developed by capitalising on the existing collaborative culture. This might be through the work of the SCA as described earlier, which has potential if supported constructively by all participants. Alternatively, there might be value in individual organisations expanding their use of formal agreements such as Memoranda of Understanding (MoUs), along the lines of the British Library, Jisc and others as described in Chapter 2. This has the benefit of strengthening strategic collaboration by setting longer term commitments and allowing individual organisations to define and position their partnership commitments effectively within their individual strategies. Whatever the routes and methods devised to progress coordinated digitisation, the key factor will be the active and ongoing commitment of the participants to support and deliver the content in such a way that the co-ordination is truly effective.

If tackling a full national digitisation strategy is too demanding, a further alternative might be to consider focusing specifically on content selection for digitisation rather than all the elements of a standard digitisation strategy as described earlier in this chapter, and to use existing outputs to define future coordination and strategy. It could be argued that while the UK has been unable to address a full strategy over the last twenty years, the country has nonetheless produced a significant amount of valuable content, out of which a *de facto* content strategy has developed organically. The broad, indicative mapping of outputs presented in Chapter 4 of this book has shown that the resources produced by the exemplar projects do not stand alone. They include some strong examples of critical mass, and also many groups of related materials which have potential to achieve critical mass over time. These are found particularly in the areas of unique and rare materials, specific themes and subjects (such as Family History, World War I, History of Medicine or Scottish and Welsh culture amongst many others) and particular formats (for example, newspapers, parliamentary publications or pamphlets). Beyond these categories, the developing work in virtual reunification and clustering also has potential to create further critical mass in the longer term.

This set of categories is by no means a coherent framework in any theoretical sense and it certainly does not cover the complete range of subjects, themes and formats that might emerge from a fuller mapping of all major UK digitisation. Nonetheless, it does suggest that there are common factors underlying existing digitised materials in the UK, and that these might be used more explicitly as a pragmatic basis for future planning.

If this approach were adopted, the collaboration of the heritage, educational and learned organisations in the UK would be needed to develop a selection framework for future digitisation. This would require an agreed, broad categorisation for digitised resources along the lines of that discussed in Chapter 4, but based on a wider range of existing outputs than it is possible to cover in this book. Each category could form a hub within the framework, onto which existing and new projects with relevant content are mapped, with links to the content itself. As new digitisation work is done, the creators could be encouraged, or preferably required by the conditions of their funding, to map their new content onto one or more of the hubs in the framework. New resources that fitted no existing hubs would suggest the introduction of a new one. The framework would therefore extend in response to content priorities as identified by digitising organisations and their audiences, rather than by a predefined model.

Such an approach could serve several functions. It could be a publicly available reference tool for the planning of further digitisation and for resource discovery at a broad level. It could allow duplication to be avoided, and identify gaps to be filled. It could highlight opportunities for further clustering, and suggest new partners for collaboration. It could also allow potential funders to consider where their support was most needed, or to be persuaded about this, as well as giving them confidence about the position and value in the national picture of the resources that they finance.

5.5 STEPPING AWAY FROM THE SILOS

This book took as its starting point the generally accepted view that the digitised content produced with public and philanthropic funding in the UK over the last 20 years has lacked effective coordination, with its selection and the resulting outputs functioning in silos and so undermining the overall value and impact of the resources created. With ever-diminishing financial support for new digitisation work, the need to achieve better coordination has become clearer still. The earlier chapters of this book show that such lack of coordination results in no small part from the complex strategic environment in the UK and changing political interests over the two decades examined. Crucially, as demand for digitised content has increased despite reduced political and financial support, the heritage, educational and learned organisations have understood the long-term role of digitisation and have remained committed to its continuation. Coordination, however, continues to be a challenge. Progress has been slow, despite a widespread desire for improvement and well-informed calls for a national digitisation strategy. This is not only a UK problem; many other nations have also grappled with the issue. Recent initiatives in some European countries and in Australia may herald progress, but their direction does not currently seem a likely course for the UK. Nonetheless, the scale and coverage of UK digitised outputs, as exemplified and discussed in Chapter 3, suggest that there has been more underlying coherence in the selection of content than might initially be thought. Identification of such coherence to support more effective planning and coordination would increase the value of past and future financial investments and of the resources produced. The UK has the advantage of a good collaborative culture amongst the organisations involved in digitisation. There would be benefit in these institutions working together to strengthen this further. The current political environment suggests that a national digitisation strategy would not be viable. Therefore, a better focus might be to consider how to improve coordination as part of the organisations' own strategic planning or to explore how past and current selection practices might be used as a basis for planning and selecting future digitised content.

There is no question that any such initiatives would require time and resources, human, technical and financial, to be scoped, implemented and maintained. For all the organisations involved in such collaboration, digitisation is only one activity within their wider range of responsibilities, and its priority will therefore always be set in that context. If improved coordination merits attention and commitment, it is because demand will not diminish, and organisations will continue to digitise to meet their users' requirements. In the current environment, the existing low level of coordination and environment of silos will simply expand. This would be unfortunate, as the UK has a substantial body of existing digitised materials, high levels of expertise in the field, a positive attitude to collaboration and much experience of cooperation. These are the building blocks for improving the coherence of the nation's digitised outputs, if the will is there to use them.

REFERENCES

British Library. (1999-2015) *Annual Report and Annual Highlights* (1999-2000 to 2014-2015). [Online] Available from: http://www.bl.uk/aboutus/annrep/index.html.

Dalbera, J-P. and Pascon, J-L. (1999) *Patrimoine culturel et multimedia*. [Paris]: Ministère de la Culture et de la Communication. [Online] Available from: http://www.culturecommunication.gouv.fr/Media/Politiques-ministerielles/Enseignement-superieur-et-recherche/Files/Numerique/Numerisation-du-patrimoine-Archives/La-politique-de-numerisation-en-France/Patrimoine-culturel-et-multimedia-Jean-Pierre-Dalbera-Jean-Louis-Pascon-novembre-1999.

European Commission. (2001) *The Lund Principles, the Lund Action Plan and its current successor*. [Online] Available from: http://cordis.europa.eu/ist/digicult/lund-principles.htm.

European Commission. (2011) *Commission Recommendation on the digitisation and online accessibility of cultural material and digital preservation*. [Online] Available from: https://ec.europa.eu/digital-agenda/sites/digital-agenda/files/en_4.pdf.

European Commission. (2015). *Digital Agenda for Europe: Member States Expert Group on digitisation & digital preservation*. Online] Available from: https://ec.europa.eu/digital-agenda/en/member-states-expert-group-digitisation-and-digital-preservation-mseg.

European Commission. (n.d.) *Digital Agenda for Europe: a Europe 2020 Initiative*. [Online] Available from: http://ec.europa.eu/digital-agenda/.

European Commission; Comité des Sages. (2011). *The New Renaissance: report of the Comité des Sages, Reflection Group on bringing Europe's cultural heritage online*. [Online] Available from: https://ec.europa.eu/digital-agenda/sites/digital-agenda/files/final_report_cds_1.pdf.

European Commission; Directorate-General for Communications Networks, Content and Technology. (2014) *Cultural Heritage: Digitisation, online accessibility and digital preservation: Report on the implementation of Commission Recommendation 2011/711/EU 2011-2013*. [Online] Available from: https://ec.europa.eu/digital-agenda/en/news/european-commissions-report-digitisation-online-accessibility-and-digital-preservation-cultural.

Follett, B.K. (2001) "Just how are we going to satisfy our research customers". *Liber Quarterly*. Vol. 11, No.3 : 218—223.

France; Ministère de la Culture et de la Communication. (2015a) Evaluation de la Politique publique de numérisation des ressources culturelles: Phase 1: Diagnostic. Paris: Ministère de la Culture et de la Communication, Secrétariat général. [Online] Available from: http://modernisation.gouv.fr/sites/default/files/epp/epp_numerisation-des-ressources-culturelles_rapport-diagnostic.pdf.

France; Ministère de la Culture et de la Communication. (2015b) *Evaluation de la politique publique de numérisation des ressources culturelles: rapport définitif Phase II*. Paris: Ministère de la Culture et de la Communication, Secrétariat général. [Online] Available from: http://www.modernisation.gouv.fr/sites/default/files/epp/epp_numérisation-des-ressources-culturelles_rapport-final.pdf.

France; Ministère de la Culture et de la Communication. (2015c) *Evaluation de la politique publique de numérisation des ressources culturelles: plan d'action*. [Online] Available from: http://www.modernisation.gouv.fr/sites/default/files/epp/epp_numérisation-des-ressources-culturelles_plan-action.pdf.

France; secrétariat général pour la modernisation de l'action publique. (2014) *Etudes sur les modèles étrangers de financement de la numérisation des actifs culturels*. [Online] Available from: http://www.modernisation.gouv.fr/sites/default/files/epp/epp_numérisation-des-ressources-culturelles_benchmark.pdf.

Government Offices of Sweden. (2011) ICT for Everyone: a digital agenda for Sweden. [Online] Available from: http://www.government.se/contentassets/8512aaa8012941deaee5cf9594e50ef4/ict-for-everyone-a-digital-agenda-for-sweden.

Green, A. (2009) 'Big digitisation: where next?'. Paper delivered at the Digital Resources for the Humanities and Arts conference, Belfast, September 2009. [Online] Available from: https://www.llgc.org.uk/fileadmin/fileadmin/docs_gwefan/amdanom_ni/dogfennaeth_gorfforaethol/darlithoedd_ac_erthyglau/dog_gorff_dar_erth_bdwn_09S.pdf.

Jisc. (2005) *Digitisation in the UK: the case for a UK framework*. [Online] Available from: http://www.webarchive.org.uk/wayback/archive/20140615225807/http://www.jisc.ac.uk/media/documents/publications/digiuk.pdf.

Jisc. (2011) *The Content Framework*. [Online] Available from: http://www.jisc.ac.uk/whatwedo/programmes/contentalliance/contentframework.aspx.

Jisc. (2012) *Strategic Content Alliance*. [Online] Available from: http://sca.jiscinvolve.org/wp/.

Mansfield, T., Winter, C., Griffith, C., Dockerty, A., and Brown, T. (2014) *Innovation Study: challenges and opportunities for Australia's Galleries, Libraries, Archives and Museums.* [No place]: Australian Centre for Broadband Innovation, CSIRO and Smart Services Co-operative Research Centre. [Online] Available from: https://sites.google.com/site/glaminnovationstudy/.

MINERVA. (2013) [Online] Available from: http://www.minervaeurope.org/.

National Audit Office. (2004) *The British Library: providing services beyond the Reading Rooms.* London: Stationery Office. HC 879 (2003-2004). [Online] Available from https://www.nao.org.uk/report/the-british-library-providing-services-beyond-the-reading-rooms/.

Norwegian Ministry of Culture. (2010) *National Strategy for Digital Preservation and Dissemination of Cultural Heritage.* Report No 24 (2008-2009) to the Storting. [Online] Available from: https://www.regjeringen.no/contentassets/f3f0e538cc704abda770db1ef2c5399b/en-gb/pdfs/stm200820090024000en_pdfs.pdf.

Norwegian Ministry of Government Administration and Reform. (c2007) *An Information Society for All.* Report No.17 (2006-2007) to the Storting. [Online] Available from: https://www.regjeringen.no/globalassets/upload/fad/vedlegg/ikt-politikk/stm17_2006-2007_eng.pdf.

Pearson, D. (2002) Digitization: do we have a strategy? *Ariadne*. Issue 30. [Online] Available from: http://www.ariadne.ac.uk/print/issue30/digilib.

Pressler, C. (2014) *National Digitisation Review: Shifting Sands: RLUK Board Briefing Paper.* [Online] Available from: http://www.rluk.ac.uk/wp-content/uploads/2014/12/RLUK-National-Digitisation-Review-CPressler.pdf.

Ross, S. and Economou, M. (1998) Information and Communications Technology in the Cultural Sector: the need for national strategies. *D-Lib Magazine*. Vol. 4, Issue 6. [Online] Available from: http://www.dlib.org/dlib/june98/06ross.html.

Ross, S., Economou, M. and Anderson, J. (1998) *Funding Information and Communications Technology in the Heritage Sector: policy recommendations to the Heritage Lottery Fund.* [Online] Available from: http://www.gla.ac.uk/schools/humanities/research/hatiiresearch/projects/completedresearchprojects/heritagelotteryfund-fundingictstudy/fundinginformationandcommunicationstechnologyintheheritagesector/.

Sero HE (2014) *Towards a UK Digital Public Space: a blueprint report.* [London]: Strategic Content Alliance. [Online] Available from: http://digitisation.jiscinvolve.org/wp/2014/12/08/towards-a-uk-digital-public-space-a-blueprint-report/.

Swedish Agency for Public Management. (2014). *Evaluation of the coordination secretariat for digitisation, digital preservation and digital accessibility (Digisam)*: 2014:16. [Online] Available from: http://www.statskontoret.se/In-English/publications/2014-summaries-of-publications/evaluation-of-the-coordination-secretariat-for-digitisation-digital-preservation-and-digital-accessibility-digisam-201416/.

Takle, M. (2009) 'The Norwegian National Digital Library'. *Ariadne*, Issue 60. [Online] Available from: http://www.mtakle.no/download/national-library.pdf.

Appendix 1

This Appendix lists the digitised outputs of the exemplar programmes and organisations discussed in the main text of this book. It is based on the information as presented on the relevant websites in 2015 (with the exception of the British Library, whose entries are taken from an internal list made available to the author). It omits outputs where no live link could be traced in May 2015.

Where the same entry appears under different organisations the duplicate title has been retained, as a record of collaborative ventures. The list excludes digitisation projects that are clearly recorded as being solely secondary sources, such as catalogues, indexes, databases and bibliographies. It also excludes projects whose purpose was to explore digitisation practices and techniques rather than content delivery.

ARCADIA FUND

African Rock Art Image Project
   http://www.arcadiafund.org.uk/grants/grant-directory/grant-details.
aspx?grantid=10112

   http://www.britishmuseum.org/research/research_projects/all_current_
projects/african_rock_art_image_project.aspx?fromShortUrl

Endangered Archives Programme
   http://eap.bl.uk/

Oral History of British Science
   http://www.bl.uk/historyofscience

ARTS AND HUMANITIES RESEARCH COUNCIL

This section covers outputs from the Council's Resource Enhancement Scheme. It is drawn from *AHRC Resource Enhancement Winners*, compiled by Alastair Dunning, Version 1, September 2009, http://web.me.com/xcia0069/ahrc.html, with links updated in May 2015 by Margaret Coutts.
   1641 Depositions
   http://www.tcd.ie/history/1641/

Acta of King Henry
http://www.rhs.ac.uk/bibl/bibwel.asp

The André Gide On-line Press Archive
http://www.gidiana.net/GA.htm

Anglia Television at the East Anglian Film Archive: a catalogue of the
collection, 1959—2000
http://www.uea.ac.uk/eafa/

The Anglo-Norman Online Hub (Phase 1)
http://www.anglo-norman.net/

The Anglo-Norman Online Hub (Phase 2)
http://www.anglo-norman.net/

Artists of Film: a database and digital archive of Arts Council-funded
documentary films 1953—1999
http://artsonfilm.wmin.ac.uk/

BBC North West Regional News and Documentary Film 1966—1985:
preservation and research access
http://ahnet2-dev.cch.kcl.ac.uk/projects/bbc_north_west_regional_
news_documentary_film_1966_1985_preservation_research_access

The Book of Curiosities: an early 11th-century Arabic cosmography
http://dhox-dev.nsms.ox.ac.uk/project/book-curiosities-early-
11th-century-arabic-cosmography

British Academic Spoken (BASE) Corpus
http://www2.warwick.ac.uk/fac/soc/celte/research/base/

The Cambridge Illuminations
http://www.fitzmuseum.cam.ac.uk/gallery/CambridgeIlluminations/

Capturing the Past Preserving the Future; digitisation of the National
Review of Live Art video collection
http://www.bris.ac.uk/theatrecollection/liveart/liveart_NRLA.html

CESAR Images: a searchable online repository of French theatre images
1600—1800
http://www.cesar.org.uk/cesar2/

CESAR, a comprehensive online repository of French Theatre resources in the 17th and 18th centuries
http://www.cesar.org.uk/cesar2/

Chopin's first editions online
http://www.cfeo.org.uk/

Classical Archaeology and Art on the Web: the Beazley Archive
http://www.beazley.ox.ac.uk/BeazleyAdmin/Script2/default.htm

The Collected Letters of Robert Southey (Parts 1—4: 1791—1815)
http://www.nottingham.ac.uk/crlc/robert-southey/

Colonial Film: moving images of the British Empire
http://www.colonialfilm.org.uk/

The Complete work of Charles Darwin Online
http://darwin-online.org.uk/

Computer Art and Technocultures: evaluating the Patric Prince Collection in the digital Age
http://www.technocultures.org.uk/

The Constance Howard Resource and Research Centre in Textiles
http://www.gold.ac.uk/textile-collection/

The Corpus of Romanesque Sculpture in Britain and Ireland
http://www.crsbi.ac.uk/

Corpus Vitrearum Medii Aevi
http://www.cvma.ac.uk/

Creation of High Wycombe Furniture Electronic Archive
http://hwfurniturearchive.bcuc.ac.uk/

The Demarco Archives: accessing a 40-year dialogue between Richard Demarco and the European Avant-Garde
http://www.vrc.dundee.ac.uk/Research/DeMarco.html

The Development of the Celtic Coin Index
http://www.celticcoins.ca/

Dictionary of Scottish Architects
http://www.scottisharchitects.org.uk/

Dictionary of the Scots Language
http://www.dsl.ac.uk/#

A Digital Edition of the Vernon Manuscript (Oxford, Bodleian Library, MS Eng.poet.a.1)
http://ahnet2dev.cch.kcl.ac.uk/projects/digital_edition_vernon_manuscript_oxford_bodleian_library_ms_engpoeta1

The Digital Image Archive of Medieval Music (DIAMM)
http://www.diamm.ac.uk/

Digital Index of Shahnama Miniature Painting
http://shahnama.caret.cam.ac.uk/

Digital Library of British Printed Images to 1700
http://www.bpi1700.org.uk/

Digitisation and Access Enhancement of the Tibetan Dunhuang Manuscripts at the British Library
http://idp.bl.uk/idp.a4d

Digitisation of Lacy's Acting Edition of Victorian Plays
http://ahnet2-dev.cch.kcl.ac.uk/projects/digitisation_lacys_acting_edition_victorian_plays

Digitisation of Renaissance Festival Books in the Collections of the British Library
http://www.bl.uk/treasures/festivalbooks/homepage.html

Digitisation of the Dictionary of the Irish language
http://dhcommons.org/projects/digitisation-dictionary-irish-language

The Digitisation of the Modern Cuttings Collection, Centre for the Study of Cartoons and Caricature
http://library.kent.ac.uk/cartoons/

Digitisation of the National Archives' Calendars of State Papers through British History Online
http://gtr.rcuk.ac.uk/project/39F3B190-0B63-451D-A65E-BDCB366CF9F4

Digitisation of the South Asian Oral History Archive
http://gtr.rcuk.ac.uk/project/41AE94B5-FEC0-4007-841B-169D1378B305

Durham Liber Vitae: A digital analysis, interlinked texts, images and research
http://www.dlv.org.uk/index.html

Early English Church Music (EECM): the fifteenth and early sixteenth centuries
http://www.eecm.net/

Early Irish Glossaries Project
http://www.asnc.cam.ac.uk/irishglossaries/

Early Stuart Libels: an electronic edition of political poems from manuscript sources
http://www.earlystuartlibels.net/htdocs/index.html

An Electronic Catalogue of Vernacular Manuscript Books of the Medieval West Midlands
http://www.mwm.bham.ac.uk/

An Electronic Corpus of 15th Century Castilian Cancionero Manuscripts; towards completion of the Dutton project
http://cancionerovirtual.liv.ac.uk/about.htm

Electronic Corpus of Lute music II
http://www.ecolm.org/

An Electronic Corpus of Medieval Welsh Prose
http://www.cardiff.ac.uk/welsh/research/projects/electroniccorpus.html

An Electronic Edition of Domesday Book (1086): interlinked translation, facsimile, databases, mapping, scholarly commentary, software
http://www.domesdaybook.net/

The Electronic Old Bailey Sessions Proceedings, c.1670—1778
http://www.oldbaileyonline.org/

An Electronic Version of Peter Clement Bartrum's Welsh Genealogies AD 300—1500
http://cadair.aber.ac.uk/dspace/handle/2160/4026

Enhancement of CASSS Digital Archive
http://dhcommons.org/projects/casss-digital-archive

An Epidoc Corpus of the Inscriptions from Aphrodisias in Caria
http://insaph.kcl.ac.uk/

EPPI: Enhanced British Parliamentary Papers on Ireland, 1801—1922
http://www.dippam.ac.uk/static_pages/eppi-archive-guide

French Interlanguage Oral Corpora
http://www.flloc.soton.ac.uk/

From Partition to Direct Rule: 50 years of Northern Ireland Parliamentary Papers online
http://stormontpapers.ahds.ac.uk/stormontpapers/index.html

Glasgow Emblem Digitisation Project
http://dhcommons.org/projects/glasgow-emblem-digitisation-project

The Henry III Fine Rolls Project
http://www.finerollshenry3.org.uk

A Historical Corpus of the Welsh Language
http://people.pwf.cam.ac.uk/dwew2/hcwl/menu.htm

Hofmeister XIX
http://www.hofmeister.rhul.ac.uk/

Imaging Papyri at Oxford
http://www.papyrology.ox.ac.uk/

The Italian Academies 1530—1650: a themed collection database
http://gtr.rcuk.ac.uk/project/A9E770E2-5DAE-4180-BCFC-9F6196
8C88B4

The Italic Epigraphy Project: text and monument
http://ahnet2-dev.cch.kcl.ac.uk/projects/italic_epigraphy_project_
text_monument

Jane Austen's Holograph Fiction Manuscripts: a digital and print resource
http://www.janeausten.ac.uk/index.html

John Ruskin's Teaching Collections
http://ruskin.ashmolean.org/collection/9006/9036/9124

The John Rylands Cairo Genizah Project
http://enriqueta.man.ac.uk/luna/servlet/ManchesterDev ∼ 95 ∼ 2

The Language of Landscape: reading the Anglo-Saxon countryside
http://www.langscape.org.uk/

Latin American Art: an online research resource
http://www.escala.org.uk/collection

The Leeds Archive of Vernacular Culture
http://www.leeds.ac.uk/english/activities/lavc/

Leeds Poetry 1950–1980
http://library.leeds.ac.uk/special-collections-leeds-poetry

Lexicon of Greek Personal Names
http://www.lgpn.ox.ac.uk/

A Linguistic Atlas of Early Middle English
http://www.lel.ed.ac.uk/ihd/laeme2/laeme2.html

A Linguistic Time-Capsule: The Newcastle electronic corpus of Tyneside English
http://www.ncl.ac.uk/necte/index.htm

The Mander & Mitchenson Theatre Collection: enhancing access for research
http://www.bristol.ac.uk/theatre-collection/explore/theatre/mander-mitchenson-collection/

Nineteenth-century Serials Edition Project (NCSE)
http://www.ncse.ac.uk/

Online Calendar of the Correspondence of Charles Darwin
https://www.darwinproject.ac.uk/calendars

The Online Froissart: a searchable electronic edition
http://gtr.rcuk.ac.uk/project/00DAC98C-1616-4390-852D-4A3361660131

Papers of Isambard Kingdom Brunel (1806–1859)
http://www.bristol.ac.uk/is/library/collections/specialcollections/archives/brunel/ikbrunel.html

The Parsed Corpus of Early English Correspondence
http://www-users.york.ac.uk/ ~lang22/PCEEC-manual/index.htm

Photographs Exhibited at the Royal Photographic Society 1870–1915
http://erps.dmu.ac.uk/

Posters of Conflict: the visual culture of public information and counter information
http://www.art.mmu.ac.uk/profile/jaulich/projectdetails/11

Primary Sources on Copyright (1450–1900) (Electronic Database of Historical Materials on Copyright from Five Key Jurisdictions)
http://www.copyrighthistory.org/

Proceedings of the Central Criminal Court 1834–1913 online
http://www.oldbaileyonline.org/

Project to Digitise the Archive of the Independent Local Radio (ILR) Programme Sharing Scheme
http://www.bournemouth.ac.uk/library/using-the-library/ilr.html

Rationalisation and Enhancement of Historic British Archaeology Collections at the Ashmolean Museum
http://www.ashmolean.org/ash/britarch/archives/archive-index.html

The Reading Experience Database 1800–1945
http://www.open.ac.uk/Arts/RED/

Remembering: Victims, Survivors and Commemoration in Post-conflict Northern Ireland
http://cain.ulst.ac.uk/victims/index.html

The Reuniting of Osip Mandelstam's Texts and Archives in Digital Form
http://digital.humanities.ox.ac.uk/project/reuniting-osip-mandelstams-texts-and-archives-digital-form

A Revised and Augmented Edition of P H Sawyer's Catalogue of Anglo-Saxon Charters
http://ahnet2-dev.cch.kcl.ac.uk/projects/revised_augmented_edition_p_h_sawyers_catalogue_anglo_saxon_charters

The St Alban's Psalter: on the Web
http://www.abdn.ac.uk/stalbanspsalter/index.shtml

Scottish Corpus of Texts and Speech (SCOTS)
http://www.scottishcorpus.ac.uk/

Scene Details in Ancient Egyptian Monuments: Oxford Expedition's electronic database and publications project (c.2960–2040 BC)
http://www.oxfordexpeditiontoegypt.com/

A Scholarly Digital Edition of Codex Sinaiticus, published on the Internet

http://www.codexsinaiticus.org/en/

Scriptorium: medieval and early modern manuscripts online
http://scriptorium.english.cam.ac.uk/

Siobhan Davies Dance Online
http://www.siobhandaviesreplay.com/

Stone in Archaeology: towards a digital resource
http://ads.ahds.ac.uk/catalogue/archive/stones_ahrb_2005/

The Survey of the Jewish Built Heritage in the United Kingdom and Ireland

http://www.jewish-heritage-uk.org/

Tibetan–Mongolian Rare Books and Manuscripts
http://www.innerasiaresearch.org/T_Msite/tmindex.html

A Trial Electronic Edition for the Early English Text Society
http://ahnet2dev.cch.kcl.ac.uk/projects/trial_electronic_edition_
preface_ancrene_wisse_early_english_text_society

TV Times Digitisation Project
http://bufvc.ac.uk/tvandradio/tvtip/

Web Access to Rock Art: the Beckensall Archive of Northumberland rock art

http://rockart.ncl.ac.uk/

A Web-mounted Database of Mid-Victorian Wood Engraved Illustration

http://www.dmvi.cf.ac.uk/

Winsor & Newton Colourman's Manuscript Archive: page-image database of historic recipes for paint making

http://www.hki.fitzmuseum.cam.ac.uk/archives/winsor-and-newton

BRITISH LIBRARY

The following is based on an internal listing of the British Library's digitisation outputs as of 2014. It covers principally digitised resources supported fully or in part by public or philanthropic funds.

19th Century British Newspapers
http://www.bl.uk/reshelp/findhelprestype/news/newspdigproj/database/

http://gdc.gale.com/products/19th-century-british-library-newspapers-part-i-and-part-ii/

Alice's Adventures in Wonderland (for Turning the Pages)
http://www.bl.uk/onlinegallery/virtualbooks.viewall/index.html#

Amaravati Sculptures
http://www.bl.uk/onlinegallery/features/amaravati/homepage.html

Ancrene Riwie
http://www.bl.uk/manuscripts/FullDisplay.aspx?ref=Cotton_MS_Cleopatra_C_VI

Anglo Saxon Charters
http://www.bl.uk/manuscripts/FullDisplay.aspx?ref=Add_MS_82931

Archival Sound Recordings
http://sounds.bl.uk

Athelstan Psalter
http://www.bl.uk/manuscripts/FullDisplay.aspx?ref=Cotton_MS_galba_a_xviii

Attributed to Joachim of Flore, Vaticinia de Pontificus
http://www.bl.uk/catalogues/illuminatedmanuscripts/record.asp?MSID=8278&CollID=8&NStart=1340

Audubon's Birds of America
http://www.bl.uk/collection-items/the-birds-of-america

Ayenbite of Inwyt
http://www.bl.uk/manuscripts/FullDisplay.aspx?ref=Arundel_MS_57

Baybars' Magnificent Qur'an (for Turning the Pages)
http://www.bl.uk/onlinegallery/virtualbooks.viewall/index.html#

Bede, Historia Ecclesiastica
http://www.bl.uk/onlinegallery/onlineex/illmanus/cottmanucoll/a/011cottiba00014u00112v00.html

Bedford Psalter and Hours
http://www.bl.uk/manuscripts/FullDisplay.aspx?ref=add_ms_42131

Beowulf
http://www.bl.uk/manuscripts/

Bible from Ethiopia (for Turning the Pages)
http://www.bl.uk/onlinegallery/virtualbooks.viewall/index.html#

Bindings
http://www.bl.uk/catalogues/bookbindings/

Blake's Notebook (for Turning the Pages)
http://www.bl.uk/onlinegallery/virtualbooks.viewall/index.html#

Bohun Psalter and Hours
http://www.bl.uk/manuscripts/FullDisplay.aspx?ref=Egerton_MS_3277

BOPCRIS (British Official Publications Collaborative Reader Information Service)
http://www.webarchive.org.uk/wayback/archive/20140615215804/
http://www.jisc.ac.uk/publications/reports/2007/18cpapersfinalreport.aspx

Botany in British India
http://www.bl.uk/reshelp/findhelpregion/asia/india/
indiaofficerecords.botanymat.html

British in India
http://www.bl.uk/reshelp/findhelpregion/asia/india/britishinindia/
index.html

Bull of Leo X, Defender of Faith
https://imagesonline.bl.uk/en/asset/show_zoom_window_popup.
html?asset=12425&location=grid&asset_list=12425&basket_item_
id=undefined

Burghley Atlas
http://www.bl.uk/onlinegallery/onlineex/unvbrit/index.html

Burney Newspapers (17th and 18th centuries)
http://find.galegroup.com/dvnw/start.do?prodId=DVNW&useGroup
Name=blibrary

Canterbury Tales
http://www.bl.uk/treasures/caxton/homepage.html

Captain R F Scott's Diary (for Turning the Pages)
http://www.bl.uk/onlinegallery/virtualbooks.viewall/index.html#

Chopin's First Editions
http://www.cfeo.org.uk/

Chopin Variorum Edition
http://www.ocve.org.uk/jsp/browse.jsp

Christine de Pizan
http://www.pizan.lib.ed.ac.uk/

Classic of Botanical Illustration (for Turning the Pages)
http://www.bl.uk/onlinegallery/virtualbooks.viewall/index.html#

Codex Palatinus Leaf
http://www.bl.uk/manuscripts/FullDisplay.aspx?ref=Add_MS_40107

Codex Sinaiticus (for Turning the Pages)
http://www.bl.uk/onlinegallery/virtualbooks.viewall/index.html#

Collect Britain (image-based collection)
http://www.bl.uk/onlinegallery

Constitution of Athens and Banks Homer: Greek Papyri
http://www.bl.uk/manuscripts/FullDisplay.aspx?ref=Papyrus_131

Cotton Manuscripts
http://www.bl.uk/reshelp/findhelprestype/manuscripts/cottonmss/
cottonmss.html

Coventry and York Doomsday
http://www.bl.uk/reshelp/findhelprestype/webres/manudigilinks/

Creton, History of Richard II
http://www.bl.uk/catalogues/illuminatedmanuscripts/record.asp?
MSID=8530&

Deed of Slave Sale
http://www.bl.uk/manuscripts/FullDisplay.aspx?ref=Papyrus_229

Dering Roll
http://www.bl.uk/manuscripts/FullDisplay.aspx?ref=Add_Roll_77720

Desert of Religion
http://www.bl.uk/onlinegallery/onlineex/illmanus/cottmanucoll/a/
011cotfaub00006u00003000.html

Diamond Sutra (for Turning the Pages)
http://www.bl.uk/onlinegallery/virtualbooks.viewall/index.html#

DigCIM (Digital Catalogue of Illuminated Manuscripts)
https://www.bl.uk/catalogues/illuminatedmanuscripts/welcome.htm

Digital Hexateuch
http://www.bl.uk/manuscripts/FullDisplay.aspx?ref=Cotton_MS_Claudius_B_IV

Digital Image Archive of Medieval Music
http://www.diamm.ac.uk/

Digital Islam Theses on EThOS
http://ethos.bl.uk

Digital Library of Hispanic Dialogues
http://www.bl.uk/reshelp/findhelprestype/webres/rarefacsimile/

Digitised Sound
http://sounds.bl.uk

Dunois Hours
http://www.bl.uk/catalogues/illuminatedmanuscripts/record.asp?MSID=6439&CollID=58&NStart=3

Durer, Drawings Add. MS 5228
https://books.google.co.uk/books?id=7Pd0qRQB18kC&pg=PA254&lpg=PA254&dq=british+library+Durer,+Drawings+Add.+MS+5228&source=bl&ots=vD2QQTf88D&sig=gWaQFFR1okry-CfsFlp6KA_uZh0&hl=en&sa=X&ei=I3hXVZCkEoHm7gabiICwCA&ved=0CCIQ6AEwAA#v=onepage&q=british%20library%20Durer%2C%20Drawings%20Add.%20MS%205228&f=false

Durer Manuscript
https://www.bl.uk/catalogues/illuminatedmanuscripts/record.asp?MSID=19008&CollID=27&NStart=39636

Durham Liber Vitae
http://www.kcl.ac.uk/humanities/cch/div/

Dutch Baroque Gardens (for Turning the Pages)
http://www.bl.uk/onlinegallery/virtualbooks.viewall/index.html#

Early English Printed Ballads
http://www.ebba.english.ucsb.edu/

Early Music Online
https://www.royalholloway.ac.uk/music/research/earlymusiconline/
home.aspx

Eighteenth Century Journals III
http://www.18thcjournals.amdigital.co.uk

EThOS (Electronic Theses Online Service)
http://ethos.bl.uk

Endangered Archives Programme
http://eap.bl.uk/

Europeana 1914–1918
http://www.europeana1914-1918.eu/

Evanion (19th-century ephemera)
http://www.bl.uk/catalogues/evanion/

Evesham Psalter
http://www.bl.uk/manuscripts/FullDisplay.aspx?ref=Add_MS_44874

Exultet Roll
http://www.bl.uk/manuscripts/FullDisplay.aspx?ref=Add_MS_30337

First Atlas of Europe (for Turning the Pages)
http://www.bl.uk/onlinegallery/virtualbooks.viewall/index.html#

Flemish Masters in Miniature (for Turning the Pages)
http://www.bl.uk/onlinegallery/virtualbooks.viewall/index.html#

Genealogical Chronicle of the English Kings
http://www.bl.uk/catalogues/illuminatedmanuscripts/record.asp?
MSID=18940

Genizah Project
http://www.genizah.org

Georgian MSS
http://www.bl.uk/reshelp/findhelplang/georgian/georgiancoll/

Ghislieri Hours
http://www.bl.uk/manuscripts/FullDisplay.aspx?ref=Yates_
Thompson_MS_29

Glimpses of Medieval Life (for Turning the Pages)
http://www.bl.uk/onlinegallery/virtualbooks.viewall/index.html#

Glorious Hebrew Prayer Book (for Turning the Pages)
http://www.bl.uk/onlinegallery/virtualbooks.viewall/index.html

Gorieston Psalter
http://www.bl.uk/manuscripts/FullDisplay.aspx?ref=add_ms_49622

Gospels of Ivan Alexander
http://www.bl.uk/onlinegallery/sacredtexts/bulggosp.html

Greek Manuscripts
http://www.bl.uk/manuscripts/

Grimbald Gospels
http://www.bl.uk/manuscripts/FullDisplay.aspx?ref=Add_MS_34890

Gutenberg Bibles
http://www.bl.uk/treasures/gutenberg/homepage.html

Handel's Messiah (for Turning the Pages)
http://www.bl.uk/onlinegallery/virtualbooks.viewall/index.html

Harley Psalter
http://www.bl.uk/catalogues/illuminatedmanuscripts/record.asp?
MSID=18402

Hebrew Manuscripts
http://www.bl.uk/reshelp/findhelplang/hebrew/manuscripts/
manuscripts.html

Henry VI Psalter
http://www.bl.uk/manuscripts/FullDisplay.aspx?ref=Cotton_MS_
Domitian_A_XVII&index=1

Henry VIII Prayer Roll
http://www.bl.uk/manuscripts/FullDisplay.aspx?ref=Add_MS_88929

Henry VIII's Psalter
http://www.bl.uk/manuscripts/FullDisplay.aspx?ref=royal_ms_2_a_
xvi

Herculaneum Papyrus
http://www.bl.uk/manuscripts/FullDisplay.aspx?ref=Papyrus_2068

Hoccleve, Regimen of Princes
http://www.bl.uk/manuscripts/FullDisplay.aspx?ref=Harley_MS_
4866

Holkham Bible Picture Book
http://www.bl.uk/onlinegallery/sacredtexts/holkham.html

Hours of Elizabeth the Queen
http://www.bl.uk/manuscripts/FullDisplay.aspx?ref=Add_MS_50001

International Dunhuang Project
http://idp.bl.uk

Iraq National Library and Archives
http://www.bl.uk/aboutus/stratpolprog/workingint/

James Gillray Satirical Prints (for Turning the Pages)
http://www.bl.uk/onlinegallery/virtualbooks.viewall/index.html#

Jane Austen
http://www.janeausten.ac.uk/index.html

Jane Austen's Early Works (for Turning the Pages)
http://www.bl.uk/onlinegallery/virtualbooks.viewall/index.html#

Jean Fouquet Miniature
http://www.bl.uk/manuscripts/FullDisplay.aspx?ref=Add_MS_37421

Jean Froissart's Chronicles
http://www.bl.uk/catalogues/illuminatedmanuscripts/record.asp?
MSID=7616&CollID=16&NStart=180501

Kathryn Marsh Collection of Children's Playground Games and Songs
http://www.wired.co.uk/news/archive/2011-03/22/british-library-
playtime

Kildare Observer Newspaper Online
http://www.kildare.ie/library/ehistory/kildare_observer_18801935/

King George III's Geographical Collections
http://www.bl.uk/onlinegallery/onlineex/kinggeorge/index.html

King George III Topographical Collection
http://www.bl.uk/onlinegallery/onlineex/kinggeorge/index.html

Know Your Place West of England
https://www.southglos.gov.uk/leisure-and-culture/tourism-and-travel/museums-and-galleries/know-your-place-west-of-england/

Lady Jane Grey's Prayerbook
http://www.bl.uk/onlinegallery/onlineex/histtexts/ladyjane/

Landmark in Medical History (for Turning the Pages)
http://www.bl.uk/onlinegallery/virtualbooks/viewall/index.html#

Leonardo's Codex Arundel (for Turning the Pages)
http://www.bl.uk/onlinegallery/virtualbooks.viewall/index.html#

Leonardo da Vinci selection (for Turning the Pages)
http://www.bl.uk/onlinegallery/virtualbooks.viewall/index.html#

Liberalism in the Americas
http://liberalism-in-americas.org/

Life of St Cuthbert
http://www.bl.uk/manuscripts/FullDisplay.aspx?ref=Yates_thompson_MS_26

Life of St Edmunds
http://www.bl.uk/manuscripts/FullDisplay.aspx?ref=Harley_MS_2278

Lindisfarne Gospels (for Turning the Pages)
http://www.bl.uk/onlinegallery/virtualbooks.viewall/index.html#

Lisbon Hebrew Bible (for Turning the Pages)
http://www.bl.uk/onlinegallery/virtualbooks.viewall/index.html#

Lorsch
http://www.bl.uk/manuscripts/FullDisplay.aspx?ref=Harley_MS_3115

Magna Carta
http://www.bl.uk/magna-carta

Malory's Morte D'Arthur
http://www.bl.uk/onlinegallery/onlineex/englit/malory/

Margery Kempe Digital Edition
http://www.bl.uk/manuscripts/FullDisplay.aspx?ref=Add_MS_61823

Marvels of the East
http://www.bl.uk/manuscripts/FullDisplay.aspx?ref=cotton_ms_vitellius_a_xv

Masterpiece of the Renaissance (for Turning the Pages)
http://www.bl.uk/onlinegallery/virtualbooks.viewall/index.html#

Medieval Fragments
http://www.bl.uk/manuscripts/5

Medieval Latin scientific manuscripts
https://www.bl.uk/manuscripts/About.aspx

Medieval Maps
http://www.bl.uk/reshelp/findhelprestype/maps/manumapcoll/mssmapsuk/findingmssmapsancient.html

Meerut Conspiracy Files, 1929–1933
http://www.britishonlinearchives.co.uk/hroup.php?pid=72696a&sid=&keywords=

Melisende Psalter
http://www.bl.uk/manuscripts/FullDisplay.aspx?ref=Egerton_MS_1139

Melrose Chronicle
http://www.bl.uk/manuscripts/FullDisplay.aspx?Source=BrowseTitles&letter=F&ref=Cotton_MS_Julius_B_XIII

Memory of the Netherlands
http://www.geheugenvannederland.nl/

Moutier Grandval Bible
http://www.bl.uk/manuscripts/FullDisplay.aspx?ref=Add_MS_10546

Motets
http://www.bl.uk/manuscripts/FullDisplay.aspx?ref=Royal_MS_8_G_VII

Mozart's Musical Diary (for Turning the Pages)
http://www.bl.uk/onlinegallery/virtualbooks.viewall/index.html#

My Ladye Nevell's Booke (for Turning the Pages)
http://www.bl.uk/onlinegallery/virtualbooks.viewall/index.html#

Nara Ehon 17th-Century Illustrated Japanese Manuscript
http://www.bl.uk/reshelp/findhelplang/japanese/japanesesection/
japancatalogues/japancatalogues.html

Nevers Gospels
http://www.bl.uk/catalogues/illuminatedmanuscripts/record.asp?
MSID=8612&CollID=8&NStart=2790

New Minster Charter
http://www.bl.uk/manuscripts/FullDisplay.aspx?ref=Cotton_MS_
vespasian_a_viii

Nineteenth-Century Serials Edition
http://www.ncse.ac.uk/index.html

Northumbrian Gospel Book
http://www.bl.uk/catalogues/illuminatedmanuscripts/ILLUMIN.
ASP?Size=mid&IllID=15071

Old Latin Genesis
http://www.bl.uk/manuscripts/FullDisplay.aspx?ref=Papyrus_2052

Oral History of British Science
http://www.bl.uk/historyofscience

Oscott Psalter
http://www.bl.uk/manuscripts/FullDisplay.aspx?ref=Add_MS_50000

Oscott Psalter Leaf
http://www.bl.uk/manuscripts/FullDisplay.aspx?ref=Add_MS_54215

Ottonian Gospels
http://en.wikipedia.org/wiki/Coronation_Gospels_(British_Library,_
Cotton_MS_Tiberius_A.ii)

Outstanding 15th-Century Church Book (for Turning the Pages)
http://www.bl.uk/onlinegallery/virtualbooks.viewall/index.html#

Pearl, Cleanness, Patience and Gawain and the Green Knight
http://britishlibrary.typepad.co.uk/digitisedmanuscripts/2012/08/sir-
gawain-and-the-green-knight-online.html

Penny Illustrated Paper
http://newspapers.bl.uk/bics/

Persian Manuscripts Digitisation
http://www.iranheritage.org/BL_Project/

Peter of Poitiers, Allegories
http://www.bl.uk/manuscripts/FullDisplay.aspx?ref=Add_MS_82947

Photographically Illustrated Books
http://www.bl.uk/catalogues/photographyinbooks/welcome.html

Picturing Canada
http://commons.wikimedia.org/wiki/Category:Images_from_the_
Canadian_Copyright_Collection_at_the_British_Library

Plant Cultures
http://www.plantcultures.org.uk

Pope, Homer
http://www.bl.uk/onlinegallery/onlineex/englit/pope/

Punch Ledgers
http://www.ljmu.ac.uk/HSS/124772.htm

Qatar Digital Library
http://www.qdl.qa/en

http://www.bl.uk/qatar/

Quadripartite indenture for Henry VII's Chapel
http://www.bl.uk/manuscripts/FullDisplay.aspx?ref=Harley_MS_
1498

Ramayana (for Turning the Pages)
http://www.bl.uk/onlinegallery/whatson/exhibitions/ramayana/
ttplaunch.html

Ramsay Psalter
http://www.bl.uk/catalogues/illuminatedmanuscripts/record.asp?
MSID=6479

Renaissance Festival Books
http://www.bl.uk/treasures/festivalbooks/homepage.html

Rinascimento Virtuale (13 Greek Palimpsests)
http://www.itz.uni-hamburg.de/RV/

RISM (Répertoire International des Sources Musicales)
http://www.rism.org.uk/

Rous Roll
http://www.bl.uk/manuscripts/FullDisplay.aspx?ref=Add_MS_48976

Royal Manuscripts Digitisation Project
http://www.bl.uk/manuscripts/

Russian Visual Arts
http://www.hri.shef.ac.uk/rva/index.html

S&A (Sisterhood and After): the Women's Liberation Oral History Project
http://www.bl.uk/learning/histcitizen/sisterhood/

St Cuthbert Gospel
http://www.bl.uk/manuscripts/FullDisplay.aspx?ref=add_ms_89000

St Omer Psalter
http://www.bl.uk/manuscripts/FullDisplay.aspx?ref=Yates_
Thompson_MS_14

La Sainte Abbaye
https://www.bl.uk/catalogues/illuminatedmanuscripts/record.asp?
MSID=18452

Shakespeare Folios
http://www.bl.uk/onlinegallery/onlineex/landprint/shakespeare/

Shakespeare Quartos
http://www.bl.uk/treasures/shakespeare/homepage.html

Ships' Logs
http://www.bl.uk/learning/langlit/changlang/writtenword/halshome/
tradingshiplogbook.html

http://www.bl.uk/reshelp/findhelpregion/asia/china/guidesources/
recordschinatrade/large14124.html

Simon Bening Calendar scenes
http://www.bl.uk/manuscripts/FullDisplay.aspx?ref=add_ms_24098

Somme le Roy
http://www.bl.uk/manuscripts/FullDisplay.aspx?ref=Add_MS_28162

Southeast Asia Digital Library
http://sea.lib.niu.edu

Spare Rib
http://www.bl.uk/spare-rib

Splendor Solis
http://www.bl.uk/manuscripts/FullDisplay.aspx?ref=Harley_MS_
3469

Stowe Breviary
http://www.bl.uk/catalogues/illuminatedmanuscripts/record.asp?
MSID=8775

Story of Public Houses
http://www.thamespilot.org.uk

Thomas Jefferson Journals
http://www.bl.uk/manuscripts

Triumphs of Charles V
https://www.bl.uk/catalogues/illuminatedmanuscripts/record.asp?
MSID=7530&CollID=27&NStart=33733

Vespasian Psalter
http://www.bl.uk/manuscripts/FullDisplay.aspx?ref=Cotton_MS_
Vespasian_A_I

Victorian Bindings
http://www.bl.uk/collections/early/victorian/bind_thu.html

Vulgate Gospels
http://www.bl.uk/onlinegallery/sacredtexts/vulgategosp.html

Wallace Correspondence Project (Natural History)
http://www.nhm.ac.uk/research-curation/scientific-resources/
collections/library-collections/wallace-letters-online/index.html?utm_
source=wallacelettersonline-short-url&utm_medium=wallacelettersonline-
short-url&utm_campaign=wallacelettersonline-short-url

Wallich Digitisation (Natural History)
http://www.kew.org/science-conservation/collections/nathaniel-
wallich/introducing-the-collection

War of 1812
http://www.bl.uk/onlinegallery/onlineex/uscivilwar/britain/
britainamericancivilwar.html

Wardington Hours
http://www.bl.uk/manuscripts/FullDisplay.aspx?ref=Add_MS_82945

Wessex Gospels
http://www.bl.uk/manuscripts/FullDisplay.aspx?ref=Royal_MS_1_
A_XIV

William Roy's Map of Scotland
http://maps.nls.uk/roy/originals.html

http://maps.nls.uk/geo/roy

Winchester Anthology
http://www.bl.uk/manuscripts/FullDisplay.aspx?ref=Add_MS_60577

Winchester Psalter
http://www.bl.uk/manuscripts/FullDisplay.aspx?ref=Cotton_MS_
nero_c_iv

Winning Endeavours: source for 20th-century international sporting
heroes
http://www.bl.uk/press-releases/2012/february/explore-londons-
olympic-history-with-winning-endeavours

World Collection Programme: India Office Digitisation Pilot
https://www.britishmuseum.org/about_us/skills-sharing/world_col-
lections_programme/british_library_in_india.aspx

Wyatt, Poems
https://imagesonline.bl.uk/?service=search&action=do_quick_
search&language=en&q=Thomas+Wyatt

Wyndham Payne Collection
http://www.bl.uk/onlinegallery/onlineex/illmanus/other/
011add000058078u00000v00.html

Yale Chaucer Project
http://www.bl.uk/treasures/caxton/homepage.html

York Mystery Plays
http://www.bl.uk/onlinegallery/onlineex/illmanus/other/
011add000035290u00004000.html

Z Safe Digitisation (high grade BL manuscripts, includes Lindisfarne Gospels)
http://www.bl.uk/manuscripts/

Zweig Phase 1(Zweig Music Manuscripts)
http://www.bl.uk/reshelp/findhelprestype/music/britishlibrary
musiconline/digitisedmusic.html

ELIB

The eLib programme focused on digitisation in the following projects:
DIAD
http://www.ukoln.ac.uk/services/eLib/projects/diad

Internet Library of Early Journals
http://www.ukoln.ac.uk/services/elib/projects/early/

The eLib programme covered other major categories (see below). Some of the individual projects listed here included moderate digitisation to support the principle outputs of the initiatives. Links to fuller information and project websites are available at http://www.ukoln.ac.uk/services/elib/projects/index.html

Hybrid Libraries

AGORA; BUILDER (Birmingham University Integrated Library Development and Electronic Resources); HEADLINE (Hybrid Electronic Access and Delivery in the Library Networked Environment); HYLIFE (Hybrid Libraries of the Future); MALIBU (Managing the Hybrid Library for the Benefit of Users)

Large-scale Resource Discovery (CLUMPS) Projects

CAIRNS (Co-operative Academic Information Retrieval Network for Scotland); M25 Link; Music Libraries Online; RIDING − Z39.50 Gateway to Yorkshire Libraries

Digital Preservation

CEDARS Project — CURL Exemplars in Digital Archives

Access to Network Resources

ADAM: Art, Design, Architecture and Media Information Gateway; Biz/
ed: Business Education on the Internet; CAIN: Conflict Archive on the
Internet; CATRIONA II; EEVL: Edinburgh Engineering Virtual Library;
IHR-Info; OMNI: Organising Medical Networked Information; ROADS:
Resource Organisation and Discovery in Subject-based services; RUDI:
Resources for Urban Design Information; SOSIG: Social Sciences Infor-
mation Gateway.

Electronic Document Delivery

EDDIS: Electronic Document Delivery; SEREN: Sharing of Educational
Resources in an Electronic Network in Wales; JEDDS: Joint Electronic
Document Delivery Software Project; LAMDA: Electronic Document
Delivery in London and Manchester; Infobike

Electronic Journals

CLIC: A parallel electronic version of an established journal — Chemical
Communications; Internet Archaeology: an international electronic journal
for archaeology; PPT: Parallel Publishing for Transactions; Superjournal
Project; Electronic Stacks Project; Electronic Seminars in History; Elec-
tronic Reviews in History; EPRESS: Electronic Publishing Resource
Service; *DeLiberations on Teaching and Learning in Higher Education*; News-
Agent for libraries: a personalised current awareness service for library and
information staff; JILT: The Journal of Information, Law and Technology;
Open Journal: the integration of electronic journals with networked in-
formation resources; Sociological Research Online; Learned Societies
Support Service.

Electronic Short Loan Projects

ACORN: Access to Course Reading via Networks; ERCOMS: Electronic
Reserve Copyright Management System; PATRON: Performing Arts
Teaching Resources Online; ResIDe: Electronic reserve for UK universities.

Images

DIGIMAP: National online Access to Ordnance Survey Digital Map Data; HELIX: Higher Education Library for Image eXchange; MIDRIB: Medical Images: Digitised Reference Information Bank.

On Demand Publishing

eOn: Electronic On Demand; Project Phoenix; Edbank; HERON; On-Demand Publishing in the Humanities (was Only Connect); SCOPE: Scottish On Demand Publishing Enterprise; ERIMS: Electronic Readings in Management Studies; Eurotext: a Collaborative Resource Bank of Learning Materials in Europe.

Pre-Prints

CogPrints: the Cognitive Sciences Eprint Archive; Education-line: Electronic Texts in Education and Training; Formations; WoPEc: Working Papers in Economics.

Quality Assurance

ESPERE: Electronic Submission and Peer Quality Review Project.

Supporting Studies

MODELS: MOving to Distributed Environments for Library Services; IMPEL2: Impact on People of Electronic Libraries: FIDDO: Focused Investigation of Document Delivery Options.

Training and Awareness

Ariadne: a parallel Web and print newsletter for librarians and information scientists; CINE: Cartoon Images for Network Education; EduLib: Educational Development for Higher Education Library staff; Netlinks: Networked Learner Support; Netskills: Network Skills Training for Users of the Electronic Library; SKIP: SKills for new Information ProfessionalS; TAPin: Training and Awareness Programme in networks.

JISC

JISC Digitisation Programme 2004—09

Phase 1 2004—07

Archival Sound Recordings
   http://sounds.bl.uk/

BL Newspapers
http://www.bl.uk/reshelp/atyourdesk/docsupply/collection/
newspapers/index.html

British Parliamentary Papers
http://www.webarchive.org.uk/wayback/archive/20140615215804/
http://www.jisc.ac.uk/publications/reports/2007/18cpapersfinalreport.
aspx

Medical Journals Backfiles
http://wellcomelibrary.org/

NewsFilm
http://bufvc.ac.uk/

Online Historical Population Reports
http://www.histpop.org

Phase 2 2007—09

19th Century Pamphlets
   http://www.britishpamphlets.org.uk

Archival Sound Recordings: British Library
http://sounds.bl.uk/

British Cartoon Archive Digitisation Project
http://www.cartoons.ac.uk

British Newspapers 1620—1900
http://www.bl.uk/reshelp/atyourdesk/docsupply/collection/
newspapers/index.html

Cabinet Papers 1914—1975
http://www.nationalarchives.gov.uk/cabinetpapers/

Digital Library of Core e-Resources on Ireland: Queen's University
Belfast
https://www.jisc-collections.ac.uk/Catalogue/FullDescription/index/
991

East London Theatre Archive
http://www.elta-project.org/home.html

Electronic Ephemera
http://johnjohnson.chadwyck.co.uk/marketing/index.jsp

First World War Poetry
http://www.oucs.ox.ac.uk/ww1lit

Freeze Frame: historic polar images: Scott Polar Research Institute
http://www.freezeframe.ac.uk

Historic Boundaries of Britain
http://www.visionofbritain.org.uk/index.jsp

Independent Radio News Archive
http://bufvc.ac.uk/tvandradio/lbc/

InViews: moving images in the public sphere
https://www.bfi.org.uk/inview/intro

Modern Welsh Journals Online
http://welshjournals.llgc.org.uk/

Pre-Raphaelite Resource Site
http://www.preraphaelites.org

UK Theses Digitisation Project
http://ethos.bl.uk/Home.do

e-Content Programme 2009–11

Connected Histories
http://www.connectedhistories.org

Mapping Crime
http://www.webarchive.org.uk/wayback/archive/20140614060821/
http://www.jisc.ac.uk/whatwedo/programmes/digitisation/econtent/
mappingcrime.aspx

Visualising China
http://visualisingchina.net

Content Programme 2011–13
Strand A
Architect US: architecturally useful scholarly resources
http://www.jisc.ac.uk/whatwedo/programmes/digitisation/
content2011_2013/ArchitectUS.aspx

CCC-EED: Context, Culture and Creativity: Enriching e-Learning in
Dance
http://www.jisc.ac.uk/whatwedo/programmes/digitisation/
content2011_2013/CCC-EED.aspx

Histology and Histopathology: virtual microscopy online
http://www.jisc.ac.uk/whatwedo/programmes/digitisation/
content2011_2013/HistologyandHistopathology.aspx

Manufacturing Pasts: industrial change in 20th-Century Britain
http://www.jisc.ac.uk/whatwedo/programmes/digitisation/
content2011_2013/Manufacturing%20Pasts.aspx

OBL4HE: Object Based Learning for Higher Education
http://www.jisc.ac.uk/whatwedo/programmes/digitisation/content2011_
2013/obl4he.aspx

Observing the 1980s
http://www.jisc.ac.uk/whatwedo/programmes/digitisation/content2011_
2013/Observing%20the%201980s.aspx

OpenLIVES: Learning Insights from the voices of Emigres from Spain
http://www.jisc.ac.uk/whatwedo/programmes/digitisation/content2011_
2013/openlives.aspx

UKVM: United Kingdom Virtual Microscope
http://www.jisc.ac.uk/whatwedo/programmes/digitisation/content2011_
2013/ukvm.aspx

Zandra Rhodes Digital Study Collection
http://www.jisc.ac.uk/whatwedo/programmes/digitisation/content2011_
2013/ZandraRhodes.aspx

Strand B
Digital Exposure of English Place-Names (DEEP)
http://www.jisc.ac.uk/whatwedo/programmes/digitisation/
content2011_2013/deep.aspx

Digitised Diseases
http://www.jisc.ac.uk/whatwedo/programmes/digitisation/content2011_2013/digitdiseases.aspx

GB/3D Fossil Types Online
http://www.jisc.ac.uk/whatwedo/programmes/digitisation/content2011_2013/gb3dfossiltypesonline.aspx

Medical Officer of Health Reports for Greater London 1848—1972
http://www.jisc.ac.uk/whatwedo/programmes/digitisation/content2011_2013/moh.aspx

Navigating 18th-Century Science and Technology: the Board of Longitude
http://www.jisc.ac.uk/whatwedo/programmes/digitisation/content2011_2013/Board%20of%20Longitude.aspx

New Connections: the BT e-Archive
http://www.jisc.ac.uk/whatwedo/programmes/digitisation/content2011_2013/newconnections.aspx

Rescue of Historical UK Sea Level Charts and Ledgers
http://www.jisc.ac.uk/whatwedo/programmes/digitisation/content2011_2013/sealevel.aspx

Rhyfl Byd 1914—1918 a'r profiad Cymreig/Welsh experience of World War One, 1914—1918
http://www.jisc.ac.uk/whatwedo/programmes/digitisation/content2011_2013/welshww1.aspx

Strand C

Integrated Broadside Ballads Archive
http://www.jisc.ac.uk/whatwedo/programmes/digitisation/content2011_2013/broadsideballads.aspx

Linking Parliamentary Records through Metadata
http://www.jisc.ac.uk/whatwedo/programmes/digitisation/content2011_2013/LIPARM.aspx

Manuscripts Online: Written Culture from 1000 to 1500
http://www.jisc.ac.uk/whatwedo/programmes/digitisation/content2011_2013/Manuscriptsonline.aspx

Old Maps Online
http://www.jisc.ac.uk/whatwedo/programmes/digitisation/content
2011_2013/Old%20Maps%20Online.aspx

Online Veterinary Anatomy Museum
http://www.jisc.ac.uk/whatwedo/programmes/digitisation/content2011_
2013/ovam.aspx

Stepping into Time
http://www.jisc.ac.uk/whatwedo/programmes/digitisation/content2011_
2013/stepping.aspx

Enriching Digital Resources 2008–09

Anglo-Saxon Cluster
http://www.ascluster.org/index.html

Automatic Biodiversity Literature Enhancement (ABLE)
http://able.myspecies.info/

Climbié Inquiry Data Corpus Online
http://www.jisc-content.ac.uk/node/60

CORRAL: UK Colonial Registers and Royal Navy Logbooks: making
the past available for the future
http://www.corral.org.uk/

Creating Heritage Artefacts for Research and Teaching in an
e-Repository (CHARTER)
https://as.exeter.ac.uk/library/news/projects/charter/

Digitisation of Countryside Images
https://www.reading.ac.uk/merl/research/merl-
digicountryimagesproject.aspx

East London Lives: a Digital Archive of 'London 2012'
https://www.jisc.ac.uk/news/olympic-archive-documents-the-
changing-face-of-east-london-30-sep-2009

Enhancing Stained Glass Studies
http://ahnet2-dev.cch.kcl.ac.uk/projects/corpus_vitrearum_medii_
aevi_phase_ii_enhancing_stained_glass_studies

Enhancing the VADS Image Collection
http://vads.ac.uk/projects/enhancingvads/

Enriching the First World War Poetry Archive
http://www.oucs.ox.ac.uk/ww1lit/

Eton Myers Collection Virtual Museum
http://www.jisc-content.ac.uk/node/286

Exposing Marandet: French Plays from the 18th and 19th Centuries
http://blogs.unimelb.edu.au/libraryintelligencer/2009/11/16/
exposing-marandet-french-plays-from-the-18th-and-19th-centuries/

Fürer-Haimendorf Archive Digitisation
https://www.soas.ac.uk/furer-haimendorf/

Historical Hansards: completing the jigsaw
http://www.kcl.ac.uk/innovation/groups/cerch/research/projects/
completed/hansards.aspx

Image Path
http://www.virtualpathology.leeds.ac.uk/jisc/

In the Bigynnyng: the Manchester Middle English digital library
http://www.library.manchester.ac.uk/inthebigynnyng/

MoDiP (Museum of Design in Plastics) Digitisation Project
http://www.modip.ac.uk/about-us/funded-projects/MoDiPDiP

Musicians of Britain and Ireland 1900–1950
http://www.peelingwall.org/uk-digitsation.html

Resurrecting the Past: Virtual Antiquities in the 19th Century
http://www.webarchive.org.uk/ukwa/target/27623467/source/alpha

The Serving Soldier
http://www.jisc-content.ac.uk/node/3

Sudan Archive Digitisation Project
https://www.dur.ac.uk/library/asc/projects/jiscsudan/

Virtual Manuscript Room
http://www.vmr.bham.ac.uk/

Welsh Ballads: completing the British Ballad Network
http://www.cf.ac.uk/insrv/libraries/scolar/digital/welshballads/insrv-scolar-welsh-ballads-jisc.html

2014

UK Medical Heritage Library
http://wellcomelibrary.org/collections/digital-collections/uk-medical-heritage-library/

NATIONAL LIBRARY OF SCOTLAND

Aberdeen Breviary
http://digital.nls.uk/aberdeen-breviary/pageturner.cfm?id=74487406

Auchinleck Manuscript
http://auchinleck.nls.uk/

Biographical Dictionary of Eminent Scotsmen
scotsmen/pageturner.cfm?id=74458002

scotsmen/pageturner.cfm?id=74458002

Blaeu Atlas of Scotland, 1654
http://maps.nls.uk/atlas/blaeu/page.cfm?seq=6

Blighty and Sea Pie
http://digital.nls.uk/blighty-and-sea-pie/pageturner.cfm?id=97133773

British Military Lists
http://digital.nls.uk/british-military-lists/pageturner.cfm?id=97343435

Chalmers' 'Caledonia'
http://digital.nls.uk/chalmers-caledonia/pageturner.cfm?id=74466563

Churchill the Evidence
http://digital.nls.uk/churchill/

The Duncan Street Explorer
http://digital.nls.uk/bartholomew/duncan-street-explorer/

Early Gaelic Book Collections
http://digital.nls.uk/early-gaelic-book-collections/pageturner.cfm?
id=75733573

English Ballads
http://digital.nls.uk/english-ballads/pageturner.cfm?id=74472158

Experiences of the Great War
http://digital.nls.uk/great-war/

First Scottish Books
http://digital.nls.uk/firstscottishbooks/

First World War 'Official Photographs'
http://digital.nls.uk/first-world-war-official-photographs/pageturner.
cfm?id=74462370

Gazetteers of Scotland, 1803–1901
http://digital.nls.uk/gazetteers-of-scotland-1803-1901/pageturner.
cfm?id=97491608

Genealogical Collections concerning Families in Scotland, made by
Walter Macfarlane,1750–1751
http://digital.nls.uk/genealogical-collections-concerning-families/
pageturner.cfm?id=74466633

Getting Started with Shakespeare
http://digital.nls.uk/shakespeare/

Golf in Scotland
http://digital.nls.uk/golf-in-scotland/index.html

A Guid Cause
http://suffragettes.nls.uk/

Gutenberg Bible
http://digital.nls.uk/gutenberg-bible/pageturner.cfm?id=74481666

Hutton Drawings
http://digital.nls.uk/hutton-drawings/pageturner.cfm?id=74466682

Jacobite Prints and Broadsides
http://digital.nls.uk/jacobite-prints-and-broadsides/pageturner.cfm?
mode=gallery&id=74466725

James VI and the Union of the Crowns
http://digital.nls.uk/unionofcrowns/

The Last Letter of Mary Queen of Scots
http://digital.nls.uk/mqs/

Map images
http://maps.nls.uk/

Medical History of British India
http://digital.nls.uk/indiapapers/background.html

Military Maps
http://maps.nls.uk/military/

Moir Rare Book Collection
http://digital.nls.uk/moir/

Morall Fabillis of Esope the Phyrgia[n]
http://digital.nls.uk/morall-fabillis-of-esope-the-phyrgian/pageturner.
cfm?id=74457640

Muriel Spark
http://digital.nls.uk/murielspark/

The Murthly Hours
http://digital.nls.uk/murthlyhours/

Northern Lights: the Scottish Enlightenment
http://enlightenment.nls.uk/

Pencils of Light: Albums of the Edinburgh Calotype Club
http://digital.nls.uk/pencilsoflight/

Phoebe Anna Traquair
http://digital.nls.uk/traquair/

Photographs of John Thomson
http://digital.nls.uk/thomson/china.html

Photographs of the South Side of Edinburgh
http://digital.nls.uk/photographs-of-the-south-side-of-edinburgh/
pageturner.cfm?mode=gallery&id=74457611

Poems Chiefly in the Scottish Dialect
http://digital.nls.uk/poems-chiefly-in-the-scottish-dialect/pageturner.
cfm?id=74464614

Propaganda — a Weapon of War
http://digital.nls.uk/propaganda/

Robert Louis Stevenson 1850—94
http://digital.nls.uk/rlstevenson/

Scotia Depicta
http://digital.nls.uk/scotia-depicta/pageturner.cfm?id=74465058

Scots Abroad: Stories of Scottish Emigration
http://digital.nls.uk/emigration/

Publications of Scottish Clubs
http://digital.nls.uk/publications-by-scottish-clubs/pageturner.cfm?id=78649753

Robert Burns 1759—1796
http://digital.nls.uk/robert-burns/

Rolls of Honour
http://digital.nls.uk/rolls-of-honour/pageturner.cfm?id=100261716

Scottish Bridges
http://digital.nls.uk/scottish-bridges/pageturner.cfm?id=74466699

Scottish Decorative Bookbinding
http://digital.nls.uk/bookbinding/1600-1.html

Scottish History in Print
http://digital.nls.uk/print/

Scottish Post Office Directories
http://digital.nls.uk/directories/

Scottish Screen Archive
http://ssa.nls.uk/

Shakespeare Collected
http://shakespeare.nls.uk/

Slezer's Scotland
http://digital.nls.uk/slezer/

Soviet Posters
http://digital.nls.uk/soviet-posters/pageturner.cfm?mode=gallery&id=74921376

Special Collections of Printed Music
http://digital.nls.uk/special-collections-of-printed-music/pageturner.cfm?id=97135480

Sporting Photos
http://digital.nls.uk/sporting-photos/category.cfm?id=13

The Spread of Scottish Printing
http://digital.nls.uk/printing/texts-titles.cfm

Theatre Posters, 1870–1900
http://digital.nls.uk/theatre-posters-1870-1900/pageturner.cfm?id=74466728

NATIONAL LIBRARY OF WALES

1588 Translation of the Bible into Welsh
https://www.llgc.org.uk/discover/digital-gallery/printedmaterial/1588welshbible/

Aberystwyth Auxiliary Temperance Society Minute Book
http://www.llgc.org.uk/collections/digital-gallery/digitalmirror-manuscripts/modern-period/temperancenlwms8323b/

Alcwyn C. Evans Pedigree Books
http://www.llgc.org.uk/collections/digital-gallery/digitalmirror-manuscripts/modern-period/alcwyn-c-evans-pedigree-books/

Alun Lewis Papers
http://www.llgc.org.uk/collections/digital-gallery/archives0/alun-lewis-papers-poetry-ms-1/

Yr Arwr, Hedd Wyn
http://www.llgc.org.uk/collections/digital-gallery/digitalmirror-manuscripts/modern-period/yr-arwr-hedd-wyn/

Autobiography of a Smuggler
http://www.llgc.org.uk/collections/digital-gallery/digitalmirror-manuscripts/modern-period/smugglersautobiographynlwms/

Beunans Ke (The Life of St Ke)
http://www.llgc.org.uk/collections/digital-gallery/digitalmirror-manuscripts/early-modern-period/beunanskenlwms23849d/

'Canu'r carchar': Prison Sonnets of T.E. Nicholas
http://www.llgc.org.uk/collections/digital-gallery/digitalmirror-
manuscripts/modern-period/t-e-nicholas/

Cardiganshire Constabulary Register of Criminals, 1897—1933
http://www.llgc.org.uk/collections/digital-gallery/digitalmirror-
manuscripts/modern-period/criminals/

Church in Wales Records
http://www.llgc.org.uk/visit/family-history/records/church-in-wales/

Collectanea Menevensia
http://www.llgc.org.uk/collections/digital-gallery/archives0/
collectaneamenevensia/

Cymru 1914: the Welsh experience of the First World War
http://cymru1914.org/en

Cynefin
http://cynefin.archiveswales.org.uk/en/tithe-maps/?utm_source=llgc_
s_f&utm_medium=ln&utm_campaign=LLGC

David Lloyd George, 1866 Diary
http://www.llgc.org.uk/collections/digital-gallery/archives0/
lloydgeorgediary/

The Diary of John Cowper Powys, 1939
http://www.llgc.org.uk/collections/digital-gallery/digitalmirror-
manuscripts/modern-period/john-cowper-powys/

Dictionary of Welsh Biography
http://yba.llgc.org.uk/en/index.html

Digital Exhibitions
http://www.llgc.org.uk/collections/digital-gallery/exhibitions/

Drawing Volumes
http://www.llgc.org.uk/collections/digital-gallery/pictures/
drawingvolumes/

Dylan Thomas, 1914—1953: Map of Llareggub (194-)
http://www.llgc.org.uk/collections/digital-gallery/digitalmirror-
manuscripts/modern-period/dylanthomasandthemapofllar/

Early Tourists
http://www.llgc.org.uk/collections/digital-gallery/digitalmirror-
manuscripts/early-modern-period/earlytouristsnlwmss22753ba/

Elis Gruffudd's Chronicle
http://www.llgc.org.uk/collections/digital-gallery/digitalmirror-manuscripts/early-modern-period/elis-gruffudds-chronicle/

Emigrants Letters
http://www.llgc.org.uk/collections/digital-gallery/digitalmirror-manuscripts/modern-period/emigrantslettersms22846d/

Etchings of Tenby
http://www.llgc.org.uk/collections/digital-gallery/pictures/etchingsoftenby/

Framed Works of Art
http://www.llgc.org.uk/collections/digital-gallery/pictures/framed-works-of-art/

Gathering the Jewels
http://www.gtj.org.uk/

Geirlyer Kyrnẁeig
http://www.llgc.org.uk/collections/digital-gallery/digitalmirror-manuscripts/early-modern-period/geirlyer-kyrn7809eig/

Geoff Charles Collection
http://geoffcharles.llgc.org.uk

Glaniad
http://www.glaniad.com/index.php?lang=en

Goronwy Owen's Cywydd Hiraeth
http://www.llgc.org.uk/collections/digital-gallery/digitalmirror-manuscripts/early-modern-period/goronwyowennlwms11568b/

History of the British Bards
http://www.llgc.org.uk/collections/digital-gallery/digitalmirror-manuscripts/early-modern-period/historyofthebritishbardsnl/

History of the Gwydir Family
http://www.llgc.org.uk/collections/digital-gallery/digitalmirror-manuscripts/early-modern-period/hostoryofthegwydirfamilynl/

Illingworth Cartoon Project
http://www.llgc.org.uk/illingworth/

John Ingleby Watercolours
http://www.llgc.org.uk/collections/digital-gallery/pictures/
inglebywatercolours/

John Thomas Photographic Collection
http://www.llgc.org.uk/collections/digital-gallery/photographs0/
johnthomas/

John 'Warwick' Smith
http://www.llgc.org.uk/collections/digital-gallery/pictures/john-
warwick-smith/

Lampeter Vestry Book
http://www.llgc.org.uk/collections/digital-gallery/archives0/
vestrybooklampeterparochial/

A Letter in the Hand of Ann Griffiths (1776—1805)
http://www.llgc.org.uk/collections/digital-gallery/digitalmirror-
manuscripts/modern-period/anngriffithsnlwms694d/

Letters from David Lloyd George to his Brother
http://www.llgc.org.uk/collections/digital-gallery/archives0/
lloydgeorgeletters/

Letters from the American Civil War
http://www.llgc.org.uk/collections/digital-gallery/digitalmirror-
manuscripts/modern-period/lettersformtheamericancivil/

Lewis Morris and William Morris' Sea Charts
http://www.llgc.org.uk/collections/digital-gallery/maps0/
lewismorrisandwilliammorris/

Liber B, John Davies of Mallwyd
http://www.llgc.org.uk/collections/digital-gallery/digitalmirror-
manuscripts/early-modern-period/liber-b-john-davies-of-mallwyd/

Llyfr Melyn Tyfrydog
http://www.llgc.org.uk/collections/digital-gallery/digitalmirror-
manuscripts/early-modern-period/llyfr-melyn-tyfrydog/

Maes y Gad i Les y Wlad/From Warfare to Welfare
http://www.myglyw.org.uk/

Morgan Llwyd, a dialogue between a child and an old man
http://www.llgc.org.uk/collections/digital-gallery/digitalmirror-
manuscripts/early-modern-period/morganllwydnlwms11431b/

The National Anthem
http://www.llgc.org.uk/collections/digital-gallery/archives0/
thenationalanthem/

Nonconformist Records
http://www.llgc.org.uk/visit/family-history/records/nonconformist/

Oxyrhynchus Papyri
http://www.llgc.org.uk/collections/digital-gallery/digitalmirror-
manuscripts/the-early-ages/the-oxyrhynchus-papyri/

Payments to a Serving Maid
http://www.llgc.org.uk/collections/digital-gallery/digitalmirror-
manuscripts/early-modern-period/paymentstoaservingmaidnlw/

People's Collection Wales
http://www.peoplescollection.wales/

Poetry by William Williams, Pantycelyn
http://www.llgc.org.uk/collections/digital-gallery/digitalmirror-
manuscripts/early-modern-period/williamspantycelynnlwms77a/

Poetry in Prose
http://www.llgc.org.uk/collections/digital-gallery/digitalmirror-
manuscripts/early-modern-period/poetry-in-prose/

Portread: Welsh Portraits Online
http://www.llgc.org.uk/collections/digital-gallery/pictures/portraits/

Prisoner of War Camp Magazines, 1943—45
http://www.llgc.org.uk/collections/digital-gallery/digitalmirror-
manuscripts/modern-period/prisoner-of-war-camp-magazines/

St Asaph Notitiae
http://www.llgc.org.uk/collections/digital-gallery/archives0/
stasaphnotitiaesamisc1300/

Salusburies of Lleweni Manuscript
http://www.llgc.org.uk/collections/digital-gallery/digitalmirror-
manuscripts/early-modern-period/salusbury/

Sir William Edmond Logan Journals, 1843—4
http://www.llgc.org.uk/collections/digital-gallery/digitalmirror-
manuscripts/modern-period/sirwilliamedmondloganjournal/

Steganographia
http://www.llgc.org.uk/collections/digital-gallery/digitalmirror-
manuscripts/early-modern-period/steganographia/

Their Past Your Future
http://www.theirpastyourfuture.org.uk/

Thomas Rowlandson
http://www.llgc.org.uk/collections/digital-gallery/pictures/
thomasrowlandson/

Thomas Taylor
http://www.llgc.org.uk/collections/digital-gallery/maps0/
thomastayloratlas5210/

A Tour in Wales
http://www.llgc.org.uk/collections/digital-gallery/pictures/
journeytosnowdon/

Tour of Hafod
http://www.llgc.org.uk/collections/digital-gallery/pictures/
tourofhafod/

Treasures Programme

Archives
http://www.llgc.org.uk/collections/digital-gallery/archives0/

Manuscripts
The Early Age
http://www.llgc.org.uk/collections/digital-gallery/digitalmirror-
manuscripts/the-early-ages/

The Middle Ages
http://www.llgc.org.uk/collections/digital-gallery/digitalmirror-
manuscripts/the-middle-ages/

Early Modern Period
http://www.llgc.org.uk/collections/digital-gallery/digitalmirror-
manuscripts/early-modern-period/

Modern Period
http://www.llgc.org.uk/collections/digital-gallery/digitalmirror-
manuscripts/modern-period/

Print Material
http://www.llgc.org.uk/collections/digital-gallery/printedmaterial/

Turner and Wales
http://www.llgc.org.uk/collections/digital-gallery/pictures/
turnerandwales/

Wales—Ohio Project
http://www.ohio.llgc.org.uk

Welsh in Patagonia
http://www.glaniad.com/

Welsh Journals Online
http://www.llgc.org.uk/welsh-journals-online/

Welsh Landscape
http://www.llgc.org.uk/collections/digital-gallery/pictures/
welshlandscape/

Welsh Newspapers Online
http://newspapers.library.wales/

Wills and Probate Records
http://www.llgc.org.uk/visit/family-history/records/wills-and-
probate/

Witchcraft in Seventeenth-century Flintshire
http://www.llgc.org.uk/collections/digital-gallery/archives0/
witchcraftcourtofgrearsessi/

Wmffre Dafis' Book of Cywyddau
http://www.llgc.org.uk/collections/digital-gallery/digitalmirror-
manuscripts/early-modern-period/wmffre-dafis-book-of-cywyddau/

Wynnstay Estate Records
http://www.llgc.org.uk/collections/digital-gallery/archives0/
stratamarcella/

Ymgyrchu! Campaign! ¡Campaña!
http://www.llgc.org.uk/ymgyrchu/index-e.htm

NATIONAL RECORDS OF SCOTLAND

Charting the Nation
http://www.chartingthenation.lib.ed.ac.uk/project.html

Register of Sasines
http://www.ros.gov.uk/services/registration/sasine-register

St Kilda and Mingulay Log Books
http://www.nas.gov.uk/about/110908.asp

Scotland's East Coast Fisheries
http://sites.scran.ac.uk/secf_final/index.php

Statistical Accounts of Scotland, 1792–1845
http://stat-acc-scot.edina.ac.uk/sas/sas.asp?action=public&passback

ScotlandsPeople

http://www.scotlandspeople.gov.uk/

This includes Statutory and Church registers; Census records; Valuation Rolls; Wills and Testaments (see following entries):

Catholic Registers: banns and marriages
http://www.scotlandspeople.gov.uk/content/help/index.aspx?r=554&1988

Catholic Registers: births and baptisms
http://www.scotlandspeople.gov.uk/content/help/index.aspx?r=554&1374

Catholic Registers: deaths and burials
http://www.scotlandspeople.gov.uk/content/help/index.aspx?r=554&1989

Catholic Registers: other events
http://www.scotlandspeople.gov.uk/content/help/index.aspx?r=554&1990

Census 1841
http://www.scotlandspeople.gov.uk/Content/Help/index.aspx?r=554&1262.
Census 1851
http://www.scotlandspeople.gov.uk/content/help/index.aspx?r=554&1261.

Census 1861
http://www.scotlandspeople.gov.uk/content/help/index.aspx?
r=554&1257.
Census 1871
http://www.scotlandspeople.gov.uk/content/help/index.aspx?
r=554&398

Census 1881
http://www.scotlandspeople.gov.uk/content/help/index.aspx?
r=554&399.
Census 1891
http://www.scotlandspeople.gov.uk/content/help/index.aspx?
r=554&400

Census 1901
http://www.scotlandspeople.gov.uk/content/help/index.aspx?
r=554&401

Census 1911
http://www.scotlandspeople.gov.uk/content/help/index.aspx?
r=554&2064

Census 1881 (LDS)
http://www.scotlandspeople.gov.uk/content/help/index.aspx?
r=554&399

Coats of Arms
http://www.scotlandspeople.gov.uk/content/help/index.aspx?
r=554&1283

Old Parish Registers: banns and marriages
http://www.scotlandspeople.gov.uk/content/help/index.aspx?
r=554&1988

Old Parish Registers: births and baptisms
http://www.scotlandspeople.gov.uk/content/help/index.aspx?
r=554&405

Old Parish Registers: deaths and burials
http://www.scotlandspeople.gov.uk/content/help/index.aspx?
r=554&1353

Soldiers Wills
http://www.scotlandspeople.gov.uk/Search/SoldiersWills/index.aspx

Statutory Registers: births
http://www.scotlandspeople.gov.uk/content/help/index.aspx?
r=554&402

Statutory Registers: deaths
http://www.scotlandspeople.gov.uk/content/help/index.aspx?
r=554&404

Statutory Registers: marriages
http://www.scotlandspeople.gov.uk/content/help/index.aspx?
r=554&403

Valuation Rolls (1875–1925)
http://www.scotlandspeople.gov.uk/content/help/index.aspx?
r=554&2080

Wills and Testaments
http://www.scotlandspeople.gov.uk/content/help/index.aspx?
r=554&407

ScotlandsPlaces

http://www.scotlandsplaces.gov.uk/

This includes Historical Tax Rolls; Ordnance Survey Name Books; RCAHMS archives; Burgh Registers; official reports; published Gazetteers and Atlases; hydrographic surveys; archaeological and architectural sites, historical maps and plans.

Scottish Archive Network

http://www.scan.org.uk/index.html

This includes historical records relating to passports; property; tax; poor relief; emigration; court proceedings; military, railway and lighthouse matters.

NOF-DIGITISE: NEW OPPORTUNITIES FUND DIGITISATION FOR LEARNING MATERIALS PROJECTS

This section is based on Education for Change. *The Fund's ICT Programmes: final evaluation report*, 2006. A more recent listing, with updated information about relevant website links, can be found at http://eprints.rclis.org/17518/.

This list was created by Alastair Dunning in 2011, and was further updated by Margaret Coutts in May 2015.

Am Baile
http://www.ambaile.org.uk

Applause South West
http://www.plymouth.gov.uk/theatrehistory

ARKive
http://www.arkive.org

Art and Architecture
http://www.artandarchitecture.org.uk/

Belfast Exposed
http://www.belfastexposed.com

BE-ME Digitise: Black&Ethnic Minority Experience
http://www.wolverhamptonart.org.uk/events/black-and-ethnic-minority-experience/

BOPCRIS
http://www.webarchive.org.uk/wayback/archive/20140615215804/
http://www.jisc.ac.uk/publications/reports/2007/18cpapersfinalreport.aspx

Brickfields
http://www.brickfields.org.uk/

British Music Information Centre Digitisation Project
http://www.bmic.co.uk

British Pathe Film Archive
http://www.britishpathe.com

Church Plans Online
http://www.churchplansonline.org/

Cistercians in Yorkshire
http://www.cistercians.shef.ac.uk/

Coalfield Web Materials
http://www.agor.org.uk/cwm/

Collect Britain
http://www.bl.uk/cbhasmoved.html

Cotton Town
http://www.cottontown.org/

Digging around Sheffield and the Peaks
http://www.idigsheffield.org.uk/

Digital Handsworth
http://www.digitalhandsworth.org.uk/

Digital Shikshapatri
http://www.shikshapatri.org.uk/

Digitised Communities Online
http://www.dshed.net/digitised

DiLoDi (Digitisation of Local Directories)
http://www.historicaldirectories.org/

Discovering Bristol: port cities
http://www.discoveringbristol.org.uk/

Dorset Coast Digital Archive
http://www.dcda.org.uk/

East of England Sense of Place: Suffolk
http://www.suffolk.gov.uk/

East of England Sense of Place: Peterborough
http://sense.peterborough.gov.uk/

Etched on Devon's Memory
http://www.devon.gov.uk/localstudies/100134/1.html

Everyday History of Tiverton and Mid-Devon
http://www.victorians.org.uk

Exploring the Potteries
http://www.exploringthepotteries.org.uk

FARNE:Folk Archive Resource North East
http://www.folknortheast.com/

Freeholders Records
http://www.proni.gov.uk/index/search_the_archives/freeholders_
records.htm

From History to Her Story: Yorkshire women's lives
http://www.historytoherstory.org.uk

From Weaver to Web
http://www.calderdale.gov.uk/wtw/index.html

The Glasgow Story
http://www.theglasgowstory.com

Greenham Common Digitisation Project
http://www.greenham-common.org.uk

Hantsphere: Hampshire's Heritage Online
http://www.hantsphere.org.uk/home

Here's History Kent
http://www.hereshistorykent.org.uk/

Heritage East Midlands a Sense of Place (HEMSOP)
http://www.hemsop.org.uk

Hidden Histories: Eastside Community Heritage
http://www.hidden-histories.org.uk/wordpress/

Hidden Lives Revealed: A Virtual Archive
http://www.hiddenlives.org.uk/

History from Headstones Online
http://www.historyfromheadstones.com/

Horseracing History Online
http://www.horseracinghistory.co.uk/hrho/action/viewHomepage

Huntley and Palmers
http://www.huntleyandpalmers.org.uk

Ideal Homes: Suburbia in Focus
http://www.ideal-homes.org.uk/

Imagine York
https://cyc.sdp.sirsidynix.net.uk/client/en_GB/yorkimages/

Ingenious
http://www.ingenious.org.uk/

inIVA Digital Archive
http://www.iniva.org

Knitting Together
http://www.knittingtogether.org.uk/

LEODIS
http://www.leodis.net/

Literary Heritage West Midlands
http://www.digitalmidlands.org.uk/lh.htm

Llechwefan (Slatesite)
http://www.llechicymru.info

London Open House (Architecture Link)
http://www.architecturelink.org.uk/homepage.html

Luxonline
http://www.luxonline.org.uk/

Music of Scott Skinner
http://www.abdn.ac.uk/scottskinner

NESBREC (North East Scotland Biodiversity Centre)
http://www.nesbrec.org.uk/

New Landscapes: enclosure in Berkshire
http://www.berkshireenclosure.org.uk/

Norfolk e-map Explorer
http://www.norfolk.gov.uk/consumption/groups/public/documents/
committee_report/jntmus150705item14pdf.pdf

Northumberland Communities
http://communities.northumberland.gov.uk

Object Lessons: using artefacts as evidence
http://www.objectlessons.org

Old Bailey Proceedings Online
http://www.oldbaileyonline.org

Past Perfect
http://www.pastperfect.org.uk/

Pathways to the Past
http://www.nationalarchives.gov.uk/pathways/default.htm

PeoplePlay UK
http://www.aboutus.org/PeoplePlayUk.org.uk

Port Cities: maritime London
http://www.portcities.org.uk/london/server/show/nav-2.html

Port Cities: Southampton: gateway to the world
http://www.plimsoll.org/

Powys: a day in the life
http://a-day-in-the-life.powys.org.uk/

Resources for Learning in Scotland
http://www.rls.org.uk/

Revolutionary Players
http://www.revolutionaryplayers.org.uk/home.stm

Royal Academy Digitisation Project
http://www.ram.ac.uk/museum/collections/academy-institutional-archives

SALIDAA (South Asian Diaspora Literature and Arts)
http://www.culture24.org.uk/am24149

Screenonline
http://www.screenonline.org.uk/

Secret Shropshire
http://www.secretshropshire.org.uk/

Side Photographic Collection
http://www.amber-online.com/gallery

SINE (Structural Images of the North East)
http://www.sine.ncl.ac.uk/

Slough History Online
http://www.sloughhistoryonline.org.uk

Spinning the Web
http://www.spinningtheweb.org.uk/

Staffordshire Past Track
http://www.staffspasttrack.org.uk

Tate Insight
http://www.tate.org.uk/

Thames Pilot
http://www.thamespilot.org.uk/

Timescapes Online
http://www.timescapes.org.uk

Transport Archives
http://www.transportarchive.org.uk/

Tudor Hackney
http://www.nationalarchives.gov.uk/education/tudorhackney/

Twixt Aire and Calder
http://www.twixtaireandcalder.org.uk/

Tyneside Life and Times
http://www.newcastle.gov.uk/leisure-libraries-and-tourism/libraries/
local-studies-and-family-history/tyneside-life-and-times

Union Makes us Strong: TUC history online
http://www.unionhistory.info/

Unnetie
https://wa5.northyorks.gov.uk/unnetie/

Victorian Times
http://www.victoriantimes.org/

Viewfinder
http://viewfinder.english-heritage.org.uk/

A Vision of Britain through time
http://www.visionofbritain.org.uk

Voice of Radicalism
http://www.abdn.ac.uk/radicalism

Window on Wiltshire's Heritage
http://www.wiltshireheritagecollections.org.uk/wiltshiresites.asp?
page=search-form

Windows on Warwickshire
http://www.windowsonwarwickshire.org.uk

NON-FORMULA FUNDING OF SPECIALISED RESEARCH COLLECTIONS IN THE HUMANITIES

Aberdeen Bestiary Project
https://www.abdn.ac.uk/bestiary/

Broadside Ballads 16th–20th Centuries (Bodleian)
http://ballads.bodleian.ox.ac.uk/

Celtic Manuscripts (Bodleian and intercollegiate)
http://image.ox.ac.uk/

English Local History Collection
http://www2.le.ac.uk/library

Gertrude Bell Archive
http://www.gerty.ncl.ac.uk/

Heritage Division: Special Collections Digitisation Programme
http://www.abdn.ac.uk/library/

Historical Manuscripts and Printed Documents Collection
http://www.library.ac.uk

Irish Collection
http://www.qub.ac.uk/directorates/InformationServices/TheLibrary/

Medieval Manuscripts
http://image.ox.ac.uk/

Mountbatten Papers
https://www.southampton.ac.uk/archives/

Papers of T.M. Johnstone
https://www.dur.ac.uk/library/

Photographic Collections
http://www.st-andrews.ac.uk/library/

Sudan Archive Cinefilms
https://www.dur.ac.uk/library/

Taylor-Schechter Genizah Collection
http://www.lib.cam.ac.uk/Taylor-Schechter/

PUBLIC RECORD OFFICE OF NORTHERN IRELAND

Allison and Cooper Collections
http://www.proni.gov.uk/index/search_the_archives/pronionflickr.
htm

First World War Journal
http://www.proni.gov.uk/index/search_the_archives/first_world_
war_journal.htm

Freeholders' Records
http://www.proni.gov.uk/index/search_the_archives/freeholders_
records.htm

Londonderry Corporation Records
http://www.proni.gov.uk/index/search_the_archives/
corporationarchive-3.htm

PRONI Records on CAIN
http://www.proni.gov.uk/index/search_the_archives/proni_records_
on_cain-2.htm

Street Directories
http://www.proni.gov.uk/index/search_the_archives/street_
directories.htm

Ulster Covenant
http://www.proni.gov.uk/index/search_the_archives/ulster_covenant.htm

Valuation Revision Book Search
http://www.proni.gov.uk/index/search_the_archives/val12b.htm

Will Calendars
http://www.proni.gov.uk/index/search_the_archives/will_calendars.htm

RESEARCH SUPPORT LIBRARIES PROGRAMME

BOOKHAD: support for nationwide research activities in the field of book
history and book design
http://www.rslp.ac.uk/projects/

BOPCRIS
http://www.webarchive.org.uk/wayback/archive/20140615215804/
http://www.jisc.ac.uk/publications/reports/2007/18cpapersfinalreport.
aspx

Charting the Nation: widening access to maps of Scotland and associated archives 1550–1740
http://www.rslp.ac.uk/projects/

CartoonHub: a national hub for British cartoons and caricature
http://www.rslp.ac.uk/projects/

Charles Booth Archive
http://booth.lse.ac.uk/

Design Council Slide Collection Cataloguing and Digitisation Project
http://www.artdes.mmu.ac.uk/visualresources/designcouncil/

The Drawn Evidence: Scotland's development through its architectural archives from industrialisation to the Millennium, 1780–2000
http://www.rslp.ac.uk/projects/

EGIL: Electronic Gateway for Icelandic Literature
http://www.egil.nottingham.ac.uk/

Glasgow Digital Library Project (The New People's Palace)
http://www.strath.ac.uk/cdlr/services/glasgowdigitallibrary/

The Visual Evidence: the photographic presentation of landscape and people
http://www.rslp.ac.uk/projects/

THE NATIONAL ARCHIVES

The resources listed in following section represent the content presented on The National Archives website under 'Online Collections' in 2015. These are accessible without charge on TNA premises. Some are subject to a charge for online access from elsewhere. Details are available on the TNA website.

Alien Arrivals
http://www.nationalarchives.gov.uk/help-with-your-research/
research-guides/alien-arrivals/

Alien Entry Books
http://www.nationalarchives.gov.uk/help-with-your-research/
research-guides/alien-entry-books/

Aliens' Registration cards 1918–1957
http://www.nationalarchives.gov.uk/records/aliens-registration-cards.
htm

Births, Marriages and Deaths in England and Wales
http://www.nationalarchives.gov.uk/help-with-your-research/
research-guides/birth-marriage-death-england-and-wales/

Bomb Sight
http://bombsight.org/#15/51.5050/-0.0900

British Army and Militia 1760–1915
http://www.nationalarchives.gov.uk/help-with-your-research/
research-guides/british-army-and-militia-1760-1915/

British Army Medal Index Cards 1914–1920
http://www.nationalarchives.gov.uk/records/medal-index-cards-ww1.
htm

British Army Nurses' Service Records 1914–1918
http://www.nationalarchives.gov.uk/records/army-nurses-service-
records.htm

British Army War Diaries 1914–1922
http://www.nationalarchives.gov.uk/records/war-diaries-ww1.htm

Cabinet Papers
http://www.nationalarchives.gov.uk/cabinetpapers/

Census Records
http://www.nationalarchives.gov.uk/records/census-records.htm

Country Court Death Duty Registers 1796–1811
http://www.nationalarchives.gov.uk/records/death-duty-registers.htm

Crime, Prisons and Punishment 1770–1935
http://www.nationalarchives.gov.uk/help-with-your-research/
research-guides/crime-prisons-punishment-1770-1935/

Death Duties 1796—1903
http://www.nationalarchives.gov.uk/help-with-your-research/
research-guides/death-duties-1796-1903/

Domesday Book
http://www.nationalarchives.gov.uk/domesday/

Durham Home Guard Records 1939—1945
http://www.nationalarchives.gov.uk/records/durham-home-guard.
htm

Famous Wills 1552—1854
http://www.nationalarchives.gov.uk/records/famous-wills.htm

First World War Soldiers' Service and Pension Records
http://www.nationalarchives.gov.uk/help-with-your-research/
research-guides/first-world-war-soldiers-pension-records/

Free Online Records: Digital Microfilm
http://www.nationalarchives.gov.uk/records/free-online-records-
digital-microfilm.htm

French Muster Rolls from the Battle of Trafalgar 1805
http://www.nationalarchives.gov.uk/records/french-muster-rolls-
trafalgar.htm

Government Datasets
http://www.nationalarchives.gov.uk/records/government-datasets.htm

Henry III Fine Rolls Project: a window into English history
1216—1272
http://www.finerollshenry3.org.uk/index.html

Household Cavalry Soldiers' Service Records 1799—1920
http://www.nationalarchives.gov.uk/records/household-cavalry-sol-
diers.htm

Irish Maps c.1558—c.1610
http://www.nationalarchives.gov.uk/help-with-your-research/
research-guides/irish-maps-c1558-c1610/

Logs and Journals of Ships of Exploration 1757—1904
http://www.nationalarchives.gov.uk/records/logs-journals-ships-of-
exploration.htm

Looted Art 1939–1961
http://www.nationalarchives.gov.uk/records/looted-art.htm

Maritime Births 1867–1960
http://www.nationalarchives.gov.uk/help-with-your-research/
research-guides/maritime-births-1867-1960/

Maritime Deaths 1781–1968
http://www.nationalarchives.gov.uk/help-with-your-research/
research-guides/maritime-deaths-1781-1968/

Maritime Marriages 1854–1972
http://www.nationalarchives.gov.uk/help-with-your-research/
research-guides/maritime-marriages-1854-1972/

Merchant Seamen Registers 1835–1857
http://www.nationalarchives.gov.uk/help-with-your-research/
research-guides/merchant-seamen-registers-1835-1857/

Merchant Seamen Registers 1918–1941
http://www.nationalarchives.gov.uk/help-with-your-research/
research-guides/merchant-seamen-registers-1918-1941/

Merchant Seamen's Campaign Medal Records 1914–1918
http://www.nationalarchives.gov.uk/records/merchant-seamen-
medals-ww1.htm

Merchant Seamen's Campaign Medal Records 1939–1945
http://www.nationalarchives.gov.uk/records/merchant-seamen-
medals-ww2.htm

Merchant Shipping Movement Cards 1939–1945
http://www.nationalarchives.gov.uk/records/merchant-navy.htm

Middlesex Military Service Appeal Tribunal 1916–1918
http://www.nationalarchives.gov.uk/help-with-your-research/
research-guides/middlesex-military-service-appeal-tribunal-1916-1918/

Naturalisation Case Papers 1801–1871
http://www.nationalarchives.gov.uk/records/naturalisation-case-
papers.htm

Non-conformist and Non-parish Births, Marriages and Deaths,
1567–1969
http://www.nationalarchives.gov.uk/help-with-your-research/
research-guides/nonconformist-non-parish-births-marriages-deaths-1567-
1969/

Passenger Lists
http://www.nationalarchives.gov.uk/help-with-your-research/
research-guides/passenger-lists/

Prisoner of War Interview Reports 1914—1918
http://www.nationalarchives.gov.uk/records/prisoners-of-war-ww1.
htm

Prisoners of War: selected records 1715—1945
http://www.nationalarchives.gov.uk/help-with-your-research/
research-guides/prisoners-of-war-selected-records-1715-1945/

Railway Employment Records 1883—1956
http://www.nationalarchives.gov.uk/help-with-your-research/
research-guides/railway-employment-records-1883-1956/

Recommendations for Military Honours and Awards 1935—1990
http://www.nationalarchives.gov.uk/records/recommendations-
honours-awards.htm

Royal Air Force Airmen Service Records 1912—1939
http://www.nationalarchives.gov.uk/help-with-your-research/
research-guides/royal-air-force-airmen-service-records-1912-1939/

Royal Air Force Combat Reports 1939—1945
http://www.nationalarchives.gov.uk/records/combat-reports-ww2.
htm

Royal Air Force Officers' Service Records 1918—1919
http://www.nationalarchives.gov.uk/records/raf-officers-ww1.htm

Royal Air Force Operations Record Books 1939—1945
http://www.nationalarchives.gov.uk/records/raf-operations-record-
books.htm

Royal Flying Corps Airmen
http://www.nationalarchives.gov.uk/help-with-your-research/
research-guides/airman-royal-flying-corps/

Royal Marines' Service Records 1842—1925
http://www.nationalarchives.gov.uk/records/royal-marines-register-
service.htm

Royal Naval Air Service Officers
http://www.nationalarchives.gov.uk/help-with-your-research/
research-guides/royal-naval-air-service-officers/

Royal Naval Air Service Officers' Service Records 1906—1918
http://www.nationalarchives.gov.uk/records/rnas-service-records.htm

Royal Naval Air Service Ratings
http://www.nationalarchives.gov.uk/help-with-your-research/
research-guides/royal-naval-air-service-ratings/

Royal Naval Division Service Records 1914—1919
http://www.nationalarchives.gov.uk/records/royal-naval-division.htm

Royal Naval Reserve Officers' Service Records 1862—1964
http://www.nationalarchives.gov.uk/help-with-your-research/
research-guides/royal-naval-reserve-officers-service-records-1862-1964/

Royal Naval Reserve Personnel
http://www.nationalarchives.gov.uk/help-with-your-research/
research-guides/royal-naval-reserve-personnel/

Royal Naval Reserve Service Records 1860—1955
http://www.nationalarchives.gov.uk/records/royal-naval-reserve-
service-records.htm

Royal Naval Volunteer Reserve Service Records 1903—1922
http://www.nationalarchives.gov.uk/records/royal-naval-volunteer-
reserve-service-records.htm

Royal Navy Officers' Service Record Cards and Files c.1840—c.1920
http://www.nationalarchives.gov.uk/records/royal-naval-officers-files.
htm

Royal Navy Officers' Service Records 1756—1931
http://www.nationalarchives.gov.uk/records/royal-naval-officers-
service-records.htm

Royal Navy Pensions 1830—1860: claims by next of kin
http://www.nationalarchives.gov.uk/help-with-your-research/
research-guides/royal-navy-pensions-1830-1860-claims-by-next-of-kin/

Royal Navy Ratings' Service Records 1853—1923
http://www.nationalarchives.gov.uk/records/royal-naval-seamen.htm

Sir Anthony Eden's Private Office Papers 1935—1946
http://www.nationalarchives.gov.uk/records/eden-papers.htm

Taxation Records
http://www.nationalarchives.gov.uk/records/government-datasets.htm

Trafalgar Ancestors Database
http://www.nationalarchives.gov.uk/trafalgarancestors/

Victoria Cross Registers 1856—1944
http://www.nationalarchives.gov.uk/records/victoria-cross-registers.
htm

Victorian Prisoners' Photograph Albums 1872—1873
http://www.nationalarchives.gov.uk/records/crime.htm

Wills 1384—1858
http://www.nationalarchives.gov.uk/records/wills.htm

Wills of Royal Navy and Royal Marines Personnel 1786—1882
http://www.nationalarchives.gov.uk/records/seamens-wills.htm

Women's Army Auxiliary Corps Service Records 1917—1920
http://www.nationalarchives.gov.uk/records/womens-army-auxiliary-
corps.htm

Women's Royal Air Force Service Records 1918—1920
http://www.nationalarchives.gov.uk/records/airwomen-ww1.htm

Women's Royal Naval Service Personnel
http://www.nationalarchives.gov.uk/help-with-your-research/
research-guides/womens-royal-naval-service-personnel/

Women's Royal Naval Service Records 1917—1919
http://www.nationalarchives.gov.uk/records/womens-royal-naval-
service-ww1.htm

World through a Lens
http://www.nationalarchives.gov.uk/through-a-lens/

WELLCOME LIBRARY

AIDS Posters
http://wellcomelibrary.org/collections/browse/collections/digaids

Arabic Manuscripts
http://wellcomelibrary.org/collections/browse/collections/digarabic

Biomedical Images
http://wellcomelibrary.org/collections/browse/collections/digbiomed

Codebreakers: makers of modern genetics
http://wellcomelibrary.org/collections/digital-collections/makers-of-modern-genetics/

Codebreakers: makers of modern genetics: digitised archives
http://wellcomelibrary.org/collections/digital-collections/makers-of-modern-genetics/digitised-archives/

Film and Sound Collection
http://wellcomelibrary.org/collections/digital-collections/film-and-sound/

Genetics Books and Archives
http://wellcomelibrary.org/collections/browse/collections/diggenetics

John Thomson Photographs
http://wellcomelibrary.org/collections/digital-collections/john-thomson-photographs/

London's Pulse: Medical Officer of Health's reports 1848—1972
http://wellcomelibrary.org/moh/

http://wellcomelibrary.org/collections/browse/collections/digmoh

Medical Journal Backfiles
http://www.webarchive.org.uk/wayback/archive/20140614054830/
http://www.jisc.ac.uk/whatwedo/programmes/digitisation/medicaljournals.aspx

Medieval and Early Modern Manuscripts
http://wellcomelibrary.org/collections/browse/collections/digwms

Mental Health Care
http://wellcomelibrary.org/collections/browse/collections/digasylum

Recipe Book Manuscripts
http://wellcomelibrary.org/collections/browse/collections/digrecipe

Recipe Books
http://wellcomelibrary.org/collections/digital-collections/recipe-
books/

Royal Army Medical Corps
http://wellcomelibrary.org/collections/digital-collections/royal-army-
medical-corps/

Royal Army Corps Medical Archives
http://wellcomelibrary.org/collections/browse/collections/digramc

Sexology
http://wellcomelibrary.org/collections/digital-collections/sexology/

http://wellcomelibrary.org/collections/browse/collections/
digsexology

UK Medical Heritage Library
http://wellcomelibrary.org/collections/browse/collections/digukmhl

http://wellcomelibrary.org/collections/browse/collections/digukmhl

Research Resources in Medical History

This programme ran from 2001 to 2012, and supported many aspects of
digital resource development. Projects including digitisation as whole or
part of the deliverables included the following:
Bethlem Royal Hospital Archives and Museum
http://archives.museumofthemind.org,uk/brha.htm

Dundee University Archives: University of Dundee Medical History
Collections
http://www.dundee.ac.uk/archives/

Florence Nightingale Museum
www.florence-nightingale.co.uk/the-collection/digitisation-
project.html
Institute of Occupational Medicine, Edinburgh: Technical Memorandum Reports
http://www.iom-world.org/

Kingston University Centre for Local Studies (part digitisation)
www.fass.kingston.ac.uk/research/historical-record/projects/hharp/

Library of the Religious Society of Friends
http://www.quaker.org.uk/TemperanceCollections

Lincolnshire Archives: Lincolnshire County Lunatic Asylum Case
Books
http://www.lincstothepast.com/exhibitions/lincolnshire-asylums-the-
treatment-of-mental-health-issues/

National Library of Scotland: official documents from Colonial India;
medical history of British India
http://www.nls.uk/

Royal College of Nursing: digitisation of The Nursing Record/British
Journal of Nursing 1888—1956
https://www.rcn.org.uk/development/library_and_heritage_services/
library_collections/rcn_archive/historical_nursing_journals

Royal Pavilion and Museums, Brighton: digitisation of historic photos
and rare periodical, Brighton and South Sussex Graphic, showing the Royal
Pavilion in use as a military hospital 1914—16
http://brightonmuseums.org.uk/

St Bartholomew's Hospital Archives and Museum: digitisation of
collection of 1826 pathological drawings c.1845—1910
http://www.agfhs.org/site/index.php/articles/78-article-2342-st-
bartholomew-s-hospital-catalogue-in-full

Appendix 2

Ayris, P. (1998) *Guidance for Selecting Materials for Digitisation.* (Joint RLG and NPO Preservation Conference: Guidelines for Digital Imaging). [Online] Available from: http://eprints.ucl.ac.uk/492/1/paul_ayris3.pdf.

Brancolini, K.R. (2000) Selecting research collections for digitization: applying the Harvard model. *Library Trends*, Vol. 48, No. 4: 783–798.

Canadian Heritage Information Network. (2012) *Capture Your Collections 2012: small museum version: bibliography.* [Online] Available from: http://www.rcip-chin.gc.ca/contenu_numerique-digital_content/numerisez_vos_collections-capture_your_collections-eng.jsp?page=bibliographie-bibliography#11.

Cornell University Library. (2004) *Cornell University Library Digital Preservation Policy Framework.* [Online] Available from: https://ecommons.library.cornell.edu/bitstream/1813/11230/1/cul-dp-framework.pdf.

Digital Library of Georgia. (2001) *Digital Library of Georgia Digitization Guide.* [Online] Available from: http://dlg.galileo.usg.edu/guide.html.

DigitalNZ. (2009a) *A Framework for Good Digitisation in New Zealand: version 2.0.* [Online] Available from: http://www.digitalnz.org/make-it-digital/selecting-for-digitisation/selection-resources.

DigitalNZ. (2009b) *Selecting for Digitisation.* [Online] Available from: http://www.digitalnz.org/make-it-digital/selecting-for-digitisation.

Federal Agencies Digitization Guidelines Initiative. (2009) *Digitization Activities: project planning and management outline.* [Online] Available from: http://www.digitizationguidelines.gov/guidelines/digitize-planning.html.

Gertz, J. (2013) Should You? May You? Can You? *Computers in Libraries*, Vol. 33, No. 2: 7–11.

Hazan, D., Horrell, J., and Merrill-Oldham, J. (1998) *Selecting Research Collections for Digitization: Full Report.* Council on Library and Information Resources. (Publication 74). [Online] Available from: http://www.clir.org/pubs/reports/hazen/pub74.html.

IFLA. (2002) *Guidelines for Digitization Projects for Collections and Holdings in the Public Domain, particularly those held by Libraries and Archives.* [Online] Available from: http://www.ifla.org/files/assets/preservation-and-conservation/publications/digitization-projects-guidelines.pdf.

JISC Digital Media. (2015) *High Level Digitisation for Audiovisual Resources.* [Online] Available from: http://jiscdigitalmedia.ac.uk/infokit/audiovisual-digitisation/audiovisual-digitisation-home.

Kenney, A.R., and Rieger, O.Y. (2001) *Report of the Digital Preservation Policy Working Group on Establishing a Central Depository for Preserving Digital Image Collections.* Cornell University Library. [Online] Available from: https://www.library.cornell.edu/preservation/IMLS/image_deposit_guidelines.pdf.

Lee, S.D. (1999) *Scoping the Future of the University of Oxford's Digital Library Collections: final report.* [Online] Available from: http://www.bodley.ox.ac.uk/scoping/report.html.

Library of Congress. (n.d.) *Preservation Digital Reformatting Program: selection criteria.* [Online] Available from: http://www.loc.gov/preservation/about/prd/presdig/presselection.html.

Lopatin, L. (2006) Library Digitization Projects, Issues and Guidelines: a survey of the literature. *Library Hi-Tech* Vol. 24, Issue 2: 273.

Minerva. (2003) *Digitisation Guidelines: a selected list*. [Online] Available from: http://www.minervaeurope.org/guidelines.htm.

National Library of Australia. (c2012) *Collection Digitisation Policy*. [Online] Available from: https://www.nla.gov.au/policy-and-planning/collection-digitisation-policy.

NEDCC. (c2007) NEDCC *Preservation Leaflets: Reformatting 6.6 Preservation and Selection for Digitization*. (Author: Janet Gertz). [Online] Available from: https://www.nedcc.org/free-resources/preservation-leaflets/6.-reformatting/6.6-preservation-and-selection-for-digitization.

NINCH. (2003) *The NINCH Guide to Good Practice in the Digital Representation and Management of Cultural Heritage Materials*. [Online] Available from: http://www.ninch.org/programs/practice.

NISO. (2007) *A Framework of Guidance for Building Good Digital Collections*. 3 ed. [Online] Available from: http://www.niso.org/publications/rp/framework3.pdf.

Ooghe, B. and Moreels, D. (2009) Analysing selection for digitisation. *D-Lib Magazine*, Vol. 15, Issue 9/10. [Online] Available from: http://www.dlib.org/dlib/september09/ooghe/09ooghe.html.

UNESCO. (n.d.) *Fundamental Principles of Digitization of Documentary Heritage*. [Online] Available from: http://www.unesco.org/new/fileadmin/MULTIMEDIA/HQ/CI/CI/pdf/mow/digitization_guidelines_for_web.pdf.

University of California Libraries. (2013) *University of California Selection Criteria for Digitization (PAG)*. [Online] Available from: http://libraries.universityofcalifornia.edu/content/university-california-selection-criteria-digitization-pag.

Yale University Library. (2006) *Report of the Working Group for Developing Selection Criteria for Collections Digitization*. [Online] Available from: http://www.library.yale.edu/iac/documents/digcriteria.final.html.

Appendix 3

MAPPING OF DIGITISATION OUTPUTS TO SELECTION CRITERIA

Tables A.1–A.3 present the totals and percentages resulting from an indicative mapping of the 741 digitisation projects and initiatives in Appendix 1 to the common selection criteria identified in Chapter 4. The mapping was based on the information available from the websites describing the projects and the content delivered. Each entry in Appendix 1 was allocated to two or more of the categories listed in the following tables as appropriate to that information.

Table A.1 Mapping of Outputs to Common Selection Criteria

Common selection criteria (see Chapter 4)	Total	Percentage (%)
Access	741	100
Value	296	40
Unique	446	60
Theme/subject	653	88
Format/medium	403	54
Coherence	53	8
Virtual reunification	7	1
Clustering	24	3

Table A.2 Mapping of Outputs to Theme or Subject categories

Theme/subject	Total	Percentage (%)
Archaeology	16	2
Asia	14	2
Canada	1	0.1
Classics	2	0.2
Genre (drama)	2	0.2
Genre (poetry)	15	2
Geography/Travel	36	5
History (all entries)	497	67
History (biographical)	17	3

Continued

Table A.2 Mapping of Outputs to Theme or Subject categories—cont'd

Theme/subject	Total	Percentage (%)
History (family)	116	16
History (gender)	3	0.4
History (general)	145	20
History (industrial)	13	2
History (local)	41	6
History (medieval)	25	3
History (military)	40	5
History (modern)	69	9
History (politics)	27	4
History (rural)	1	0.1
Iceland	1	0.1
Ireland	16	2
Language (all entries)	52	7
Language (Arabic)	1	0.1
Language (English)	21	3
Language (French)	5	0.7
Language (Gaelic)	2	0.2
Languages (general)	4	0.5
Language (Greek)	4	0.5
Language (Hebrew)	1	0.1
Language (Irish)	2	0.2
Language (Scots)	3	0.4
Language (Welsh)	9	1
Law	1	0.1
Literature and Literary Studies (all entries)	92	12
Literature (English)	31	4
Literature (French)	4	0.5
Literature (Gaelic)	2	0.2
Literature (general)	31	4
Literature (Greek)	4	0.5
Literature (Hebrew)	2	0.2
Literature (Latin)	2	0.2
Literature (Russian)	1	0.1
Literature (Scots)	3	0.4
Literature (Spanish)	1	0.1
Literature (Welsh)	11	1
Middle East	6	0.8
Music	29	4
Performing arts	13	2
Philosophy	11	1
Religion/theology	77	10
Russia	2	0.2
Science (all entries)	60	8
Science (biosciences/evolution)	19	3

Table A.2 Mapping of Outputs to Theme or Subject categories—cont'd

Theme/subject	Total	Percentage (%)
Science (earth)	1	0.1
Science (engineering/computing)	2	0.2
Science (general)	7	0.9
Science (horticulture)	2	0.2
Science (medicine)	22	3
Science (nutrition)	3	0.4
Science (physics)	3	0.4
Science (veterinary)	1	0.1
Scotland	63	9
Society/Culture	84	11
Sport	9	1
United States of America	1	0.1
Visual arts	87	12
Wales	48	6

Table A.3 Mapping of Outputs to Format or Medium categories

Format/medium	Total	Percentage (%)
Bookbindings	3	0.4
Film	26	4
Images	90	12
Journals/serials	10	1
Maps	19	2
Newspapers	6	0.7
Palimpsests	1	0.1
Pamphlets	1	0.1
Parliamentary papers	2	0.2
Posters	1	0.1
Sound/audio	13	2
Theses	2	0.2

INDEX

Zeitfracht Medien GmbH
Ferdinand-Jühlke-Straße 7
99095 Erfurt, Deutschland
produktsicherheit@kolibri360.de